PROCEEDINGS OF SPIE

SPIE—The International Society for Optical Engineering

Industrial Sensing Systems

Anbo Wang
Eric Udd
Chairs/Editors

5–6 November 2000
Boston, USA

Sponsored and Published by
SPIE—The International Society for Optical Engineering

SPIE

P

PROCEEDINGS
SERIES

Volume 4202

SPIE is an international technical society dedicated to advancing engineering and scientific
applications of optical, photonic, imaging, electronic, and optoelectronic technologies.

The papers appearing in this book compose the proceedings of the technical conference cited on the cover and title page of this volume. They reflect the authors' opinions and are published as presented, in the interests of timely dissemination. Their inclusion in this publication does not necessarily constitute endorsement by the editors or by SPIE. Papers were selected by the conference program committee to be presented in oral or poster format, and were subject to review by volume editors or program committees.

Please use the following format to cite material from this book:

Author(s), "Title of paper," in *Industrial Sensing Systems*, Anbo Wang, Eric Udd, Editors, Proceedings of SPIE Vol. 4202, page numbers (2000).

ISSN 0277-786X
ISBN 0-8194-3867-7

Published by
SPIE—The International Society for Optical Engineering
P.O. Box 10, Bellingham, Washington 98227-0010 USA
Telephone 1 360/676-3290 (Pacific Time) • Fax 1 360/647-1445
http://www.spie.org/

Printed in the United States of America.

Contents

Conference Committee

Conference Chairs

Anbo Wang, Virginia Polytechnic Institute and State University (USA)
Eric Udd, Blue Road Research (USA)

Program Committee

John W. Berthold III, McDermott Technology Inc. (USA)
Daniele Inaudi, SMARTEC SA (Switzerland)
Alexis Mendez, ABB (USA)
Ralf Pechstedt, Bookham Technology Ltd. (UK)
Robert S. Rogowski, NASA Langley Research Center (USA)
Robert J. Schroeder, Schlumberger-Doll Research (USA)
Whitten L. Schulz, Blue Road Research (USA)
Daniel B. Thompson, Air Force Research Laboratory (USA)
Marek T. Wlodarczyk, OPTRAND, Inc. (USA)

Session Chairs

Fiber Optic Sensors for Automotive Applications
Marek T. Wlodarczyk, OPTRAND, Inc. (USA)
Ralf Pechstedt, Bookham Technology Ltd. (UK)

Aerospace Applications
Daniel B. Thompson, Air Force Research Laboratory (USA)
Robert S. Rogowski, NASA Langley Research Center (USA)

Oil and Gas Applications
Robert J. Schroeder, Schlumberger-Doll Research (USA)
Alexis Mendez, ABB (USA)

Smart Civil Structures
Daniele Inaudi, SMARTEC SA (Switzerland)
Whitten L. Schulz, Blue Road Research (USA)

Utilities and Power Generation
John W. Berthold III, McDermott Technology Inc. (USA)

Prospects for practical and economic FOS solutions for automotive measurement applications*

Peter McGeehin**

Compton Consultants, School Road, Compton, Newbury, RG20 6QU, UK

ABSTRACT

The advancement of fibre optics in sensing has been inhibited in the last two decades by the lack of availability of suitable components. Whilst transduction principles have been fully explored, and some key applications (for example fibre optic gyroscopes) have moved forward, even the up-take of the ubiquitous fibre Bragg gratings as sensors is being inhibited by both practical and cost considerations. Optical sensors, it seems, can be created, but the means of extracting the information from them, building on the precision and systems advantages that fibre optics should bestow, has proved difficult. If sources, detectors and essential signal processing can be combined in such a way that, essentially, the benefits that fibre optics bestow in telecoms networks can be harnessed for sensing, then fibre optic sensing can truly be said to have realised its promise. Integrated optics modules employing silicon as both substrate and waveguide are being developed and deployed rapidly for telecommunications applications, through which they will become cost effective quite quickly. They have also been demonstrated as useful modules when employed in sensing systems as, for example, Sagnac and white light interferometers. In the latter form they have been employed to process signals in an optical system measuring the cylinder pressure in a four-cylinder internal combustion engine. It is argued that this specific application, demanding as it is, illustrates the generic potential of such modules for broad application in many sensing situations.

Keywords: Optical fibre sensors, automotive, applications, silicon, integrated, fiberoptics, waveguides, pressure

1. INTRODUCTION

This paper describes an integrated optic signal processing approach to a generic problem, that of diaphragm deflection in pressure transducers. It does so by describing a possible solution to an application that presents fibre optic sensing with a considerable engineering and signal processing challenge. The application is critical path in character and has not been accomplished successfully in production using non-optical measurement technologies. If achieved successfully, that is in both technical and cost terms, it will indeed be possible to say that, finally, the application of fibre optics in sensing has achieved it's potential. The application is the direct measurement, in real time, of an in-cylinder pressure sensor for automotive production engine management applications. The application demands that the system follows the pressure transient in each firing cylinder and undertakes analysis of this signal in order to establish control parameters for engine fuelling, implemented through the engine management system.

1.1   The application

Pressure measurement is becoming increasingly important in vehicle management and control systems. Legislation to reduce engine emissions to the minimum remains strong, with perhaps the greater effort now being applied to diesel engines. Improved engine management systems require sensors that can provide more direct guidance on engine conditions than the panoply of more remote sensors presently employed. Such model based control systems need direct information about engine parameters in real time. The pressure profile in individual cylinders is one potential source of such information. The utility of this measurement is demonstrated by the importance attached to it during engine development.

*This paper presents the personal opinions of the author.
**Peter McGeehin, is also with the Department of Systems Engineering, Brunel University, Uxbridge UB8 3PH, UK, and can be contacted at pmcg@compton-consultants.co.uk, phone +44 1635 578 455, fax +44 1635 578 016.

Modified spark plugs each incorporating a pressure sensor have been used, but their design can introduce spurious 'organ pipe' effects. Besides determination of knock, maximum cylinder pressure and maximum pressure rise, comparisons can be performed between different pressure curves in order to optimise efficiency. Dithering techniques can be employed to adjust spark advance. The heat balance in the engine can be enhanced by adjusting the cooling system. Operation of spark ignition direct injection engines is much improved by cylinder pressure measurement together with comparison between the pressure transients in both inlet and exhaust manifolds.

In general the analysis of cylinder pressure yields the following information:

> Peak pressure and the maximum pressure gradient are related to the strength of materials of construction and therefore the weight of the engine (more useful as development tools).

> The indicated mean effective pressure is equivalent to the internal energy and is therefore a criterion for power output and mechanical (friction) losses.

> The pumping mean effective pressure is a criterion for the effectiveness of gas exchange, influenced by the inlet and exhaust valves and ports.

> Burn duration is a criterion for efficiency and fuel consumption.

> The shape of the heat release is a criterion for emissions, noise and efficiency.

> Cylinder pressure indicates the regularity of combustion, thus allowing ignition failures to be identified as a maintenance diagnostic or knock (which can be both noisy and destructive) to be avoided operationally.

Cylinder pressure measurement systems generally use piezoelectric high temperature pressure sensors, high insulation cables, charge amplifiers, a crank angle sensors, A-to-D converters and fast data processing for signal analysis and display. Engine management systems have been developed for normally operating large bore gas engines used for gas compression and other applications, here the strain gauge sensor (Kistler C-sensor) is mounted in a modified pressure indicator or safety valve attachment. In a production vehicle engine environment, real improvements in driving characteristics and efficiency optimisation have been demonstrated (first experiences being in motor sport).

1.2   Manifold injected engines

In Europe, there is almost no interest in in-cylinder pressure sensing in the arena of manifold injected gasoline engines. The current sensing technology which is utilized is sufficient to operate the manifold injected gasoline engine properly. North American OBD II fuel economy demands may lead to the need for in-cylinder pressure sensing in order to identify misfiring of a cylinder which may not be detected by ion-stream measurement via the spark plug. Other than this, there does not seems to be an advantage from using in-cylinder pressure sensors for this type of engine. It is not however clear as to whether the OBD II requirements can be fulfilled by the methods presently under development.

1.3   Gasoline (GDI) and diesel (DDI) direct injection engines

In these engines the fuel is injected, with precision in both timing and quantity, directly into the combustion region of the cylinder. In Europe, engineers working in the GDI and DDI arena are quite interested in having a cylinder pressure signal available for real-time closed loop engine control. Presently, only secondary sensor input sources are being used to get a rough idea about the combustion pressure and its time evolution during the combustion cycle. Pressure information is the first hand information needed to control the combustion process. So far no reliable and cost effective technology has become available, and this seems to have inhibited the adoption of DI technology.

It is very difficult to make a statement about the timing of the introduction of such a pressure sensor engine management system. The reliability has to be proven in a number of durability tests, the design being required to survive the life of an engine (approx. 6000 hours of operation, $\sim 5 \times 10^8$ cycles). The design has to be extremely robust to survive the high pressure (150-200 bar) in a diesel engine and the high temperatures (up to 900°C) experienced particularly in turbo charged petrol engines. It should withstand thermal shock (maximum 0.20% drift). The accuracy must also be sufficient in the low pressure range (ambient pressure) to allow an observation of the compression curve. The market remains unconvinced about robustness and reliability during real time durability tests. Also, there would need to be changes in the electronic control unit (ECU) hardware and its algorithm. A universal list of advantages for the technology is anticipated (savings in fuel consumption, output power increase, better NVH, less emission etc.) but it is difficult to obtain quantitative information on the improvements that can be made.

There remain, however, some other practical difficulties. The first is sensor location. There is reluctance to allow an additional bore in the cylinder head on account of cost and space constraints. Integration of the pressure sensor into the spark plug, glow plug or injection nozzle is favoured. The second is the target price. There is general reluctance to introduce improvements at extra cost, but in the case of GDI engines there appears to be a willingness to pay a premium for this technology compared to the current costs of knock-sensors and ion-stream measurement (which could be eliminated by in-cylinder pressure sensing). Even in this case the target system price is challenging: ~ $5-$8/cylinder would be acceptable! Though the pressure on manufacturers is such that systems might be sold in tens of million of units per year, this does represent a considerable challenge. Application of the technology in engine development and test work represents a higher priced route for market entry. The heavy duty truck application also offers substantial potential (and perhaps another, higher priced entry into the market by which the technology is perfected).

2. SIGNAL PROCESSING TECHNOLOGY

The use of silicon as an integrated optics substrate has been realised in a slightly unexpected way. Silicon itself is just sufficiently transparent at 1300-1600 nm to make it viable as the waveguiding medium in small devices. ASOC [®] [(1)] technology embraces several distinct elements building elegantly upon a non-optical, indeed micro-electronic, substrate platform. The conventional VLSI wafer comprises a top layer of silicon (0.1-0.2 μm thick) separated from the silicon substrate by a buried layer of silicon oxide ~0.4 μm thick, created, for example, by ion implantation. The top layer of silicon is increased to a thickness suitable for a ridge waveguide by, for example, epitaxial growth. Once at a suitable thickness this layer can then be patterned to create ridge waveguides, wells and alignment features for light sources, detectors and optical fibres. Methods for creating features in the layer include ion milling, and chemical etching (for example using potassium hydroxide). The silicon adjacent to features can be doped by diffuson, or ion implantation, and electrodes (for example made from evaporated aluminium) deposited so that the light propagating down the waveguide can be modulated electronically. In Figure 1 the structure of an ASOC[®] waveguide is illustrated. This technology makes extensive use of silicon microengineering principles and can be conveniently classified, with slightly fuzzy boundaries, as follows [(2)] :

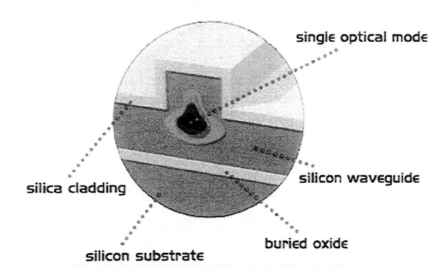

Figure 1: Structure of a ridge waveguide made on a silicon substrate (courtesy Bookham Technology)

a. Manufacturing methods

>   Fibre attachment and alignment
>   Packaging, especially sealing the fibre with strain relief
>   Attenuation measurement and adjustment
>   Methods for attaching and aligning light source and detector to the waveguide substrate

b. Passive components

>   Multiplexer/demultiplexer (WDM)
>   Waveguide taper and coupling structures (suitable perhaps for a multi-mode-fibre to mono-mode-waveguide converter)
>   Circuit board connecting links
>   Absorbers for stray and guided light
>   Transceiver structure

c. Active components and modulators

>   Using thermo-optic activation or isolation methods
>   Electro-optic switch and cross-over
>   Phase modulator
>   Polariser

Thus far, telecoms component product families incorporating leading edge sources and detectors have been launched [3] with significant impact on the market, including transceiver and electronically variable attenuators. Sensors and sensor modules have been created in development using this process knowledge, including a fibre optic gyroscope transmitter/receiver, a Mach-Zehnder interferometer (demonstrated for pressure measurement), a temperature stable interferometer and a 'white light' interferometer

STANDARD WHITE LIGHT INTERFEROMETER

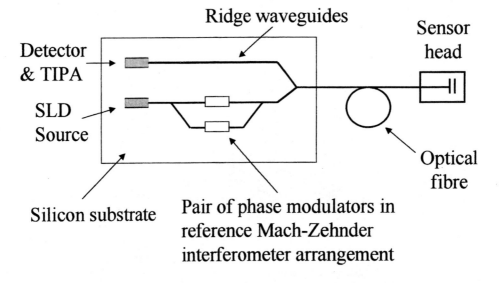

Figure 2: Optical schematic of what might be considered a 'standardised' white light interferometer platform. Packaged the device might be 50mm x 25 mm x 12 mm without fibre pigtail. Several interferometers could be integrated on a single chip

This is quite clearly a significant technology. Its relevance to optical sensing is obvious and can be illustrated by reference to WLI. Figure 2 illustrates what might be regarded as a 'standard' WLI platform. The output from the SLD is split and fed through a pair of diffusion doped phase modulators in the path-imbalanced Mach-Zehnder reference interferometer (MZI), which introduces a modulated optical path difference (OPD). The interfering light beams are fed by single mode optical fibre to the sensor head that is typically a Bragg grating, Fabry-Perot or Fizeau cavity with OPD matched to that of the MZI. Reflected light is guided back to the InGaAs PIN detector, the silicon-based trans-impedance amplifier providing a convenient electronic output. In operation the maximum transducer displacement is typically half the source peak wavelength and the module can be operated at the quadrature point of the two complementary outputs. ASOC® technology enables the integration of miniature semiconductor light source chips onto the silicon chip containing the photonic circuit. In

Figure 3: Illustrating the integrated optic chip, on the left a laser chip aligned with the waveguide, (middle) an aligned fibre and (right) showing the scale of these devices.

a similar fashion, photodiodes can be integrated onto the same silicon chip. This is a distinct advantage in comparison to discrete systems and other optical technologies, resulting in a compact and rugged system with the potential for automated high-volume manufacture. Figure 3 shows photographs of components on chip. Phase modulation of the optical waveguide modes can be achieved by thermal means or via a pin-junction formed by a controlled doping process. By applying a current to the pin-junction, electrons and holes will be injected into the light path. This causes a change in refractive index of the waveguide at the point of injection and hence, varies the speed of the light beam propagating through the phase modulator which is equivalent to a modulation of its phase. The phase modulators are used in the pressure sensor system enabling signal processing to linearise the inherently sinusoidal response of an interferometric sensor and their function in the sensor system is explained in more detail later.

3. APPLICATION FOR AUTOMOTIVE ENGINE MANAGEMENT

For optimal functioning of the white-light pressure sensor system, the optical path difference 2d of the sensor head and the path imbalance of the Mach-Zehnder interferometer (MZI) must be matched within a few microns. The photolithographic masks employed to produce the photonic structures on the ASOC silicon chip guarantee the required repeatability of the path imbalance of the MZI. On the sensor head side, a silicon diaphragm based optical cavity can be used as a pressure sensitive element for in-cylinder measurements. A silicon diaphragm assembly can be mounted into different styles of adapters to fit different packages as needed. Further, by changing the diaphragm thickness and/or its size, the sensor can be adapted to cover the required pressure range.Using a slightly more complex photonic layout on the ASOC signal processing chip, several pressure sensor heads can be interrogated with one single chip. The demonstrated sensing principle is generic and can be used to measure a number of different physical quantities by replacing the pressure sensor head with a sensor head whose optical path difference varies with the quantity to be measured.

Apart from these principle advantages a number of additional benefits related to manufacture can be derived. The signal processing chip analysing the optical return signal from the sensor head is based on ASOC technology, lending itself to low-cost, high-volume manufacture through standard silicon processing on wafer scale. The ability to hybridise photodiodes and light sources in chip form results in enormous cost savings in comparison to a system realised with discrete components. This lends itself to a more compact and robust design. Passive alignment methods for light source and optical fiber attachment will further reduce volume assembly costs. The pressure sensing diaphragm can be manufactured using standard silicon processing on wafer scale. This combines the excellent mechanical properties of silicon with a low-cost, high-volume manufacturing approach. Other manufacturing methods and materials could be employed.

The system has been installed into a direct injection diesel engine for analysis [4]. The DI diesel installation provides the harsh real environment that in-cylinder pressure sensors need to survive for production application. Engine mounted sensors are subject to the high engine vibration levels, as well as the high in-cylinder pressures and temperatures. The ability to operate reliably on a diesel platform readily supports the demands of other engine platforms, specifically direct injection gasoline. The performance objective is to provide comparable signal performance to instrument grade sensors from a production durable solution.

Figure 4 shows some initial signal performance results from the diesel test engine compared to a Kistler 6123 engine test transducer. Figure 4a shows the signal from a Kistler (upper trace) and the optical sensor (lower trace) from the same cylinder. Plot (b) shows the same signals overlayed showing the main region of error being the low pressure region. Plot (c) shows the in-cylinder linearity and hysteresis vs. the Kistler, with plot (d) showing the overall pressure error over a number of complete cycles. The system shows extremely encouraging signal characteristics over the pressure range, with slight drift in the low pressure region. These results are from prototype sensors that have not been subject to any refinement or thermal drift optimisation. Engine control requirements use both the high and low pressure characteristics of the engine cycle and it is recognised that this low pressure capability significantly enhances the total system capability and the overall cost effectiveness of using in-cylinder pressure sensors. Further refinement of the sensor is needed to improve the low range performance. Similarly, the low pressure drift is the primary cause of variance with the reference sensor. Improving this will greatly increase the system accuracy.

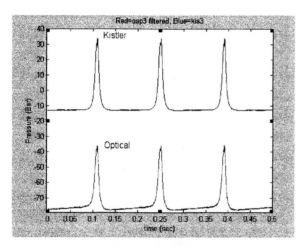

Figure 4a: Kistler and optical pressure trace

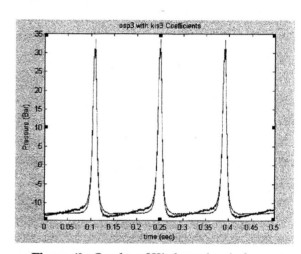

Figure 4b: Overlay of Kistler and optical sensor

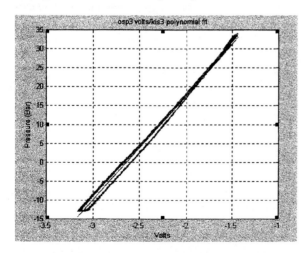

Figure 4c: Output characteristic

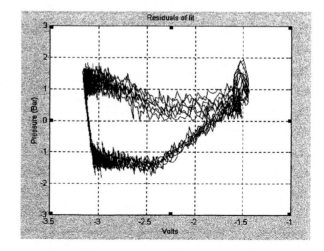

Figure 4d: Overall error over many cycles

When the Kistler signal and the new pressure sensor signal from the same cylinder, are processed using the same IMEP calculation, similar results are achieved as shown in Figure 5. IMEP is one of the parameters that can be calculated from the in-cylinder pressure signal and is core to a number of engine control functions.

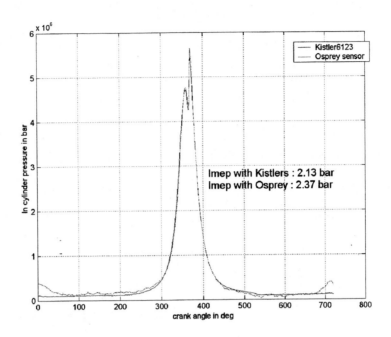

Figure 5. IMEP calculation results ('Osprey' is the project name for the silicon optoelectronic solution)

CONCLUSION

This paper has addressed an application that presents a fibre optic sensing signal processing system with a considerable engineering challenge. The application is the direct measurement, in real time, of an in-cylinder pressure sensor for automotive production engine management applications. It demands that the system follows the pressure transient in each firing cylinder and facilitates analysis of this signal in order to establish control parameters for engine fuelling through the engine management system. This application is critical path in character and has not been accomplished successfully on production using non-optical measurement technologies. If achieved successfully, that is in both technical and cost terms, it will indeed be possible to say that, finally, the application of fibre optics in sensing has achieved it's potential.

A fibre optics-compatible integrated optic signal processing system employing, for illustration, a conventional micro-engineered Fabry-Perot diaphragm sensor, has been used to demonstrate an in-cylinder pressure transduction system which has the potential for high volume, low cost production. Cylinder pressure within the combustion chamber of a 4-cylinder direct injection diesel engine has been measured and good comparability achieved with a conventional engine test transducer.

The advancement of fibre optics in sensing has been inhibited in the last two decades by the lack of availability of suitable components. Whilst transduction principles have been fully explored, and some key applications (for example fibre optic gyroscopes) have moved forward, even the up-take of the ubiquitous fibre Bragg gratings as sensors is being inhibited by both practical and cost considerations. Optical sensors, it seems, can be created, but the means of extracting the information from them, building on the precision and systems advantages that fibre optics should bestow, has proved difficult. If sources, detectors and essential signal processing can be combined in such a way that, essentially, the benefits that fibre optics bestow in telecoms networks can be harnessed for sensing, then fibre optic sensing can truly be said to have realised its promise. Integrated optics modules employing silicon as both substrate and waveguide are being developed and deployed rapidly for telecommunications applications, through which they will become cost effective quite quickly. Their use as white light interferometers to process signals in an optical system measuring the cylinder pressure in a four-cylinder internal combustion engine illustrates the generic potential of such modules for broad application in many sensing situations.

REFERENCES

1. ASOC® is a Registered Trademark of Bookham Technology. The company was a member of OSCA for a period during the 1990's.
2. See the following patents and their equivalents: Manufacturing methods: WO9835253, WO9743676, WO9927403, WO9963373, WO9839677, GB2339300, GB2330424, WO9742534, WO9957591, WO9961946, WO0002079, WO9966360, WO0007247, WO0010045. Modulators and active components: WO9957600, WO0010039, WO0004416, US5908305, WO9508787, GB2265252, WO9912062, WO9921036, WO0024867, GB2340616, GB2339919, GB2339301,GB2333851. Passive components: WO9857205, WO9934539, WO9928772, GB2334395, GB2334396, WO9835250, WO9960433. Sensors and sensor modules: WO9915856, WO9822775, WO9966358, WO9960341
3. See www.bookham.com/products/
4. M. Fitzpatrick, R. Pechstedt and Yicheng Lu, *SAE paper OOP-162* (1998)

Slow wave structures application for oil monitoring

Andrey A. Yelizarov[*]

Moscow State Institute of Electronics & Mathematics
3/12 Bolshoj Trechsvjatitelskyi Lane, Moscow, 109028, Russia.

ABSTRACT

This paper presents new measuring technology based on the application of Slow-Wave Structures (SWS). The use of SWS for RF-measurement transducers makes it possible quite efficiently to monitor and measure parameters of various industial processes and materials [1-2]. Difficulties arise when monitoring and measuring parameters of oil-liquids such as the conductivity or the continuity of flow. This is caused by the strong screening action of the monitored oil-liquid on electromagnetic field of the slowed wave in the helical sensitive elements (HSE). As the conductivity increases, the screening becoms even stronger, and this leads to a reduction in the sensitivity and measurement accuracy.
The solution of such problems requires the creation and modeling of more complex desings of sensitive elements in which the screening action is weakened. Such an effect can be achieved by matching the field of a hybrid slowed wave to the medium being monitored, utilizing HSE based on layered magnetic and dielectric structures with a smooth variation of the electrodynamic parameters.

Keywords : Slow Wave Structure, Helical Sensitive Element, Oil Monitoring.

1. STARTING RELATIONS

Let us consider a model for a sensitive element in the form of a cylindrical helix, wound on a thin walled dielectric tube (Fig.1). The helix is bounded by a coaxial cylindrical shield whose conductivity we will assume to be ideal. The inner radius of the dielectric tube is a, the average radius of the helix is b, the inner radius of the shield is d. Let us number the regions, starting from the inner region (1, 2, 3), using in the following text numbers of the regions as the subscripts on the quantities corresponding to these regions.

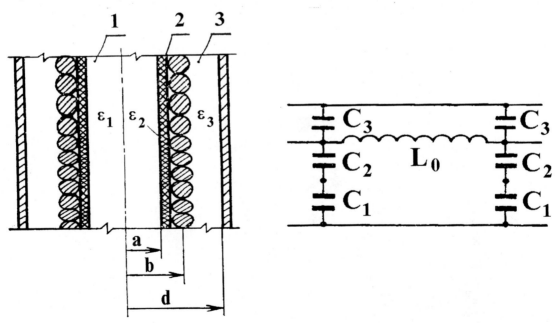

Fig.1 Model of sensitive element in the form of a celindrical helix, wound on a thinwalled tube made of dielectric.

Fig.2 Circuit diagram for three-conductor equivalent line replacing the model of helical sensitive element.

When SWS application for oil monitoring, in particular a HSE, it is important to correctly choose the operating frequency. For too high a frequency, the field of the wave is clamped to the surface of the SWS and interaction with the oil-medium proves to be ineffective. For too low a frequency, the field is concentrated in the region between the SWS and the shield. In the case under consideration this leads to a decrease in the effectiveness of the interaction between the slow electromagnetic wave and the oil-medium inside the dielectric tube. However, due to the low dielectric constant of oil and the high dielectric constant of water, even at relatively low frequencies a significant part of the energy of the electric field remains within the HSE, distributed in the oil-medium and the dielectric tube. This allows us to make use of approximations for analysis of the HSE under consideration, significantly simplifying its calculation.

Let us replace the HSE with a tree-conductor equivalent line (Fig.2); the middle conductor replaces the SWS proper, the lowest conductor replaces the regions inside the SWS, the upper conductor replaces the regions outside the SWS. The equivalent line capacitance C_1 characterizes the capacitance inside the dielectric tube; C_2, between the dielectric tube and the helix; C_3, between the helix and the shield; L_0 is the equivalent line inductance of the structure.

Let us assume that the components of the field depend on the time t and coordinate z in the direction of propagation of the wave as $\exp(j\omega t - j\beta z)$, where ω is the angular frequency, β is the phase constant.

Using the relatively low frequency approximation, we determine the capacitances C_2 and C_3 as the capacitances between the coaxial cylinders

$$C_2 \approx 2\pi\varepsilon_0\varepsilon_2 / [\ln(b/a)],$$

$$C_3 \approx 2\pi\varepsilon_0\varepsilon_3 / [\ln(b/a)] \tag{1}$$

where ε_0 is the dielectric constant of free space; ε_2 and ε_3 are the relative dielectric constants of the tube material and the oil-medium in the region between the helix and the shield respectively.

We determine the capacitance C_1 in the terms of the electrical conductivity Y^e on the inner surface of the dielectric tube [3]:

$$C_1 = -2\pi a \tau_1^2 Y^e / j\omega, \tag{2}$$

where τ_1 is the transverse constant connected with the phase constant β by the relation

$$\tau_n^2 = \beta^2 - \omega^2\varepsilon_n\mu_0.$$

Here n is the number of the regions; μ_0 is the magnetic permeability of free space.

Making use of the expression for Y^e in the absence of an internal boundary for the region [3],

$$Y^e = -j\omega\varepsilon_1 I_1(a\tau) / \tau_1 I_0(a\tau),$$

where ε_1 is the complex dielectric constant of the monitored oil-medium; I_0, I_1 are modified Bessel functions of the first kind, using (2) we obtain

$$C_1 = 2\pi a \tau_1 \varepsilon_0 \varepsilon_1 I_1(a\tau_1) / I_0(a\tau_1). \tag{3}$$

For relatively low frequencies, when the arguments of the Bessel functions are considerably less than unity, we can use the following approximate substitution [4]:

$$I_1(a\tau_1) \approx a\tau_1 / 2; \qquad I_0(a\tau_1) \approx 1. \tag{4}$$

This allows us to write instead of (3)

$$C_1 \approx 2\pi(a\tau_1)^2 \varepsilon_0\varepsilon_1 / 2.$$

The line inductance L_0 of a cylindrical helix with a shield is determined by following expression:

$$L_0 = \frac{\beta^2\mu_0 I_1(b\tau)K_1(b\tau)}{2\pi\tau^2}\{1 - \frac{I_1(b\tau)K_1(d\tau)}{K_1(b\tau)I_1(d\tau)}\}\frac{(2\pi b)^2}{h^2}, \qquad (5)$$

where h is the step of the helix; K_1 is a modified Bessel function of the second kind; τ is the transverse constant in the regions adjacent to the helix.

In the considered case, the value of the transverse constants in the regions adjacent to the helix are not identical. However, in the cases of practical interest, when the slow coefficient is rather long, the ratio β^2/τ^2 can be set to be equal to unity.

In the case of relatively low frequencies, when $b\tau, d\tau << 1$, along with the approximations in (4) we can use the approximations

$$K_1(b\tau) \approx 1/b\tau; \qquad K_1(d\tau) \approx 1/d\tau,$$

as a result of which expression (5) is considerably simplified

$$L_0 \approx \frac{\pi b^2}{h^2}\{1 - b^2/d^2\}.$$

2. CALCULATION OF THE ATTENUATION COEFFICIENT

In order to estimate the sensitivity of the measurements, we can use the interaction coefficient K_i introduced in [5]. Characterizing the ratio of the energy of the electric field of the wave in the region of the monitored oil-medium to the total of the wave, the interaction coefficient enters as a factor into the expression for the power loss P_{loss} per unit length of the system

$$P_{loss} = 2{,}1\cdot 10^{-8} fnK_i P\varepsilon'' / \varepsilon', \qquad (6)$$

where f is the frequency, Hz; n is the slow coefficient; P is the transmitted power; ε'' is the imaginary part and ε' is the real part of the relative dielectric constant of the monitored material or medium.

The ratio P_{loss}/P determines the wave attenuation coefficient K_a (dB), which is an informative parameter in measuring low conductivities

$$K_a = 4{,}34\cdot P_{loss} l/P, \qquad (7)$$

where l is the length of the SWS, m.

The slow coefficient n is determined in terms of the parameters of the equivalent line according to the formula

$$n = \sqrt{L_0 C_0 / \varepsilon_0\mu_0},$$

where C_0 is the overall line capacitance of the SWS, which can be determined by replacing the series-connected capacitances C_1 and C_2 with a single capacitance and combining it with the capacitance C_3. As the result we obtain

$$C_0 \approx \varepsilon_1 2\pi(a\tau_1)^2\{1 - \frac{(a\tau)^2(b-a)}{2b} + \frac{2\pi\varepsilon_2}{\ln(d/b)}\}.$$

The ratio $\varepsilon'' / \varepsilon'$ appearing in (6) for oil with water medium is determined by the expression

$$\varepsilon'' / \varepsilon' = \sigma / (8l\varepsilon_0 2\pi f) - 0{,}22\sigma \cdot 10^9 / f , \qquad (8)$$

where σ is the conductivity of oil-medium.

Solving (6), (7), and (8) simultaneously, we obtain

$$K_a \approx 20 \cdot nK_i \sigma l. \qquad (9)$$

From (9) we see that the attenuation coefficient does not depend directly on frequency and for a specified value of the conductivity is proportional to the product of the interaction coefficient on the slow coefficient. For the slow coefficient on the order of 10...100, $K_i > 0{,}1$ and length of the HSE on the order of 0,1 m, we obtain $K_a > (2...20)\sigma$, which makes measurement for values of σ greater than 0,1S/m realistic. For low conductivities, either the length of the measured section or the slow coefficient should increase.

We can increase the slow coefficient by using coupled (coaxial) helices with opposing windings (Fig.3). For distances between helices considerably shother than their radii and relatively low frequencies, when the product of the length of the helical loop on the slow coefficient is cosiderably less than the free-space wavelength

$$n \approx 2\varepsilon_{rel}(b+a)^2 / h(b-a), \qquad (10)$$

where a and b are the radii of the inner and outher helices respectively; ε_{rel} is the relative dielectric constant of the oil-medium between the helices.

If the monitored oil-medium passes trough the small gap between the helices, then we can concentrate almost the entire electric field in the liquid, increase the slow coefficient, and thus increase the value of K_i.

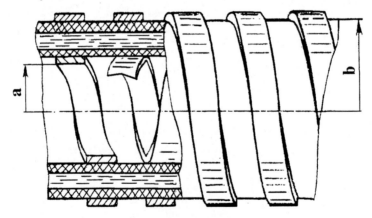

Fig.3 Model of sensitive element in the form of coupled (coaxial) helices with opposing windings.

3. OPTIMIZATION OF THE PARAMETERS OF THE HELICAL SENSITIVE ELEMENTS

From the relations given above, it follows that optimization of the parameters of HSE is reduced to obtaining the maximum value of the product nK_i. Let us determine the interaction coefficient for a helix wound on a dielectric tube filled with the monitored oil-medium.

Obviously K_i will be higher as the thickness of the tube becomes smaller. This allows us, by choosing $(b-a) << b$, to write instead of (1)

$$C_i \approx \varepsilon_0 \varepsilon 2\pi b / (b - a),$$

Using the technique outlined in [5] and setting $\tau_1 \approx \beta$ in the case of rather high slow coefficient, for the interaction coefficient in the considered model we obtain

$$K_i = \frac{\tau_0^2}{2\pi\beta_0^2} \frac{\varepsilon_1}{\left\{\dfrac{a\tau_1\varepsilon_1\varepsilon_2 I_1(a\tau_1)}{\varepsilon_1\tau_1 I_1(a\tau_1)(b-a)+\varepsilon_2 I_0(a\tau_1)} + b\tau_1\dfrac{K_1}{K_0}(b\tau_1)\dfrac{\left\{1+\dfrac{I_1}{K_1}(b\tau_1)\dfrac{K_0}{I_0}(d\tau_1)\right\}}{\left\{1-\dfrac{I_0}{K_0}(b\tau_1)\dfrac{K_0}{I_0}(d\tau_1)\right\}}\right\}}, \quad (11)$$

where τ_0 and β_0 are the transverse and phase constants for the monitored oil-medium.

The relation (11) allows us to optimize the parameters of the HSE, but the need to ensure a sufficiently uniform distribution of the electric field energy of the wave, determining the efficiency of the HSE, restricts the parameter $a\tau_1$ to a relatively small value for which the longitudinal electric field depends little on the radius. Analysis shows that the parameter $a\tau_1$ should not exceed 0,8 for $b\tau_1$=0,9.

In Fig.4, we show the dependenes of the interaction coefficient on the ratio $\varepsilon_2 / \varepsilon_1$ as the values of d/b vary from 1,2 to 2,0; relative dielectric constant of the oil-medium is ε_1 =2,0; $a\tau_1$=0,8; $b\tau_1$=0,9.

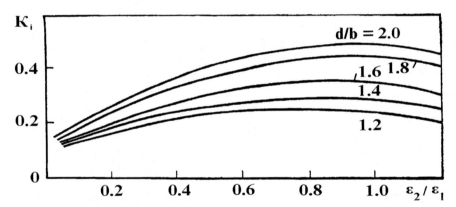

Fig.4 Dependences of the interaction coefficient on the ratio $\varepsilon_2 / \varepsilon_1$.

From the curves obtained we see that K_i increases with an increase in the distance from the helix to the shield and for close values of the relative dielectric constants for the tube material ε_2 and the monitored oil ε_1.

The calculation of the interaction coefficient is considerably simplified in the case of coupled helices. Earlier it was noted that at relatively low frequencies, practically all the electric field energy is concentrated between the helices. If the relative dielectric constant of the material insulating the helix from the oil-medium is taken to be equal to the real part of the relative dielectric constant of oil, and the distribution of the electric field energy over the entire region between the helices will be uniform and the ratio of the fraction of the energy in the oil to the entire electric field energy will be equal to the ratio of the thickness of the oil layer W to the gap between the helices $(b - a)$. Since in helical SWS the electromagnetic energy is distributed between waves of the electric and magnetic types and mainly the electric type wave interacts with the oil, the maximum possible value of K_i cannot be greater than 0,5. Thus

$$K_i \approx 0,5W / (b-a). \quad (12)$$

4. CONCLUSION

Thus our analysis of HSE has shown that it is possible in principle to design simple measuring devices for monitoring contamination of oil accompanied by the appearance of conductivity on the order of 10^{-2} S/m or more. In this case, the operating frequencies of the primary measuring transducers can be chosen in the range 10-1000 MHz, which allows us to use standart analog-to-digital transducers and microprocessors for treatment of the signal.

REFERENCES

1. Yu.N.Pchel'nikov and A.A. Yelizarov, "Technological sensors on slow electromagnetic waves", Proceedings SPIE Microelectronics Manufacturing Symposium. Vol. *"Process. Equipment and Materials Control in Integrated Circuit Manufacturing IV"*. Santa Clara, CA, USA, #3705-25, pp.225-236, 1998.
2. Yu.N.Pchel'nikov and A.A.Yelizarov, "Nondestructive testing of semiconductors and thin coatings", Proceedings SPIE Microelectronics Manufacturing Symposium. Vol. *"Process, Equipment and Materials Control in Integrated Circuit Manufacturing"*. Austin, TX, USA, #2876-18, pp.180-188, 1996.
3. L.N. Loshakov and Yu.N.Pchel'nikov, *Theory and Design of Traveling Wave Tube Amplification* [in Russian], Sovetskoe Radio, Moscow, 1964.
4. A. Grey and G.B. Mathews, *A Treatise on Bessel Functions and Their applications in Physics,* 2nd ed., Macmillan, London, 1931.
5. A.A.Yelizarov, "Design and application of the engineering method for calculating the efficiency of interaction of slowed electromagnetic waves with dielectric and semiconducting media". Author's Abstract, Moscow,1993.

----

* Correspondence : Email:lmis@miem.edu.ru; Telephone: 095 916 1216.

Study on the Composite Cure Process Optimization
With the Aid of Fiber Optic Sensors

Boming Zhang   Zhanjun Wu   Dianfu Wang

Center for Composite Materials
Harbin Institute of Technology, Harbin, 150001, P. R. China

ABSTRACT

In order to improve the mechanic performance, the cure process mechanics are approached to find out the relationships among the viscosity history, void evolution and the ultimate mechanic property of the composite structure. A computer simulation program, on the base of resin flow model, thermal stress model, material model and void model, is developed to simulate the cure process, with which the cure process parameter is optimized to meet this challenge. The information of cure process, which mainly is the viscosity evolution and gel point, is obtained by fiber optic microbend sensor made specially. The reason to sensing and controlling the viscosity evolution is that it can represent the cure degree to some extent, besides, keep the viscosity at the lowest point as long as possible can reduce the void and keep appropriate fiber content in the ultimate composite structure. The data is processed by an expert system based on the computer simulation program, then the expert system give instruction to the temperature controller and pressure controller to regulate the cure process.

Keywords: cure process, void, viscosity, fiber optic sensor

1. INTRODUCTION

Advanced composite has been utilized broadly and greatly in aerospace, military and civil industry because of its high specific strength and stiffness. However, the process technology is still not mature compare to those traditional materials such as steel and polymers. For advanced composite, it is difficult to insure the quality of composites for the reason that there are complicated processes during curing process of composites. The cure cycle has a significant effect on the quality of the finished part. The traditional cure cycle based on empirical approach could not ensure the quality of cured products on the aspects of unstable performance, the high cost of production and the low efficiency.

There are mainly two methodologies adopted to improve the processing technology: one is theory analysis, which includes process mechanic models setting up based on actual cure process, Springer, Tsai, Gutowski and some other scientists have done lots of outstanding works, they set up a series of models which are very close to actual process, such as thermal—chemical principle、the resin flow and fiber function principle、the void action principle and the residual stress principle. Once the models are set up, computer simulation became possible and practicable. The other approach is knowledge-based expert system control of the cure process, at the same time, with the aid of an in-situ monitoring system. There are many methods to realize in-situ monitoring, but, in my opinion, the fiber optic sensors monitoring is the most promising way since its low susceptibility to electromagnetic interference and some other ideal features. In the 1990s, numerous papers and reports have involved into fiber optic sensors. According to the reports, there are mainly three kinds of fiber optic sensors used. They are based on evanescent wave spectroscopy, refractive index change and near-infrared spectroscopy respectively. In this presentation, a very new fiber optic sensor is developed and utilized to monitor the cure process of the advanced fiber reinforced resin matrix composite.

2. THE NEW FIBER OPTIC SENSOR

The new type of fiber optic sensor is designed on the base of microbend principle. In optical fiber, transmitted light intensity is lost in response to microbending. This loss occurs most frequently when the highest-order mode in the fiber core is coupled to the first cladding (radiation) mode, which then is rapidly attenuated. Mechanical coupling of structural changes to an embedded fiber modifies the fiber geometry and allows light to be coupled between modes, thus modulating the light intensity transmitted by the fiber. Fibers in smart structures may be embedded within composites such as carbon epoxy. These materials have carbon reinforcing fibers in an epoxy matrix where the periodicity is greatly diminished. A spatial periodicity may be present in the event that a prepreg is used. Thus elements of intrinsic microbend sensing may be employed in smart structure applications[2,3]. Experimental setup for composite cure monitoring is shown in Fig. 1. Sensor considerations can be divided into two general areas, the fiber microbending sensitivity, represented by $\Delta T/\Delta X$, and the mechanical design of the device, associated with $\Delta X/\Delta F$

$$\frac{\Delta T}{\Delta F} = \frac{\Delta T}{\Delta X}\frac{\Delta X}{\Delta F}$$

where T is the fiber transmission, F is the applied force, and X the deformation of fiber normal to its axis. Sensitivity to microbending, or large $\Delta T/\Delta X$, is increased by small values of numerical aperture. It has been found that the microbend sensitivity of telecommunication optical fibers is low. One example of the many influences on the total sensitivity of the microbend sensor is given by the length dependence of microbend sensitivity. It has been found that the microbend sensitivity of the sensor

$$\frac{\Delta T}{\Delta X} \propto l^q$$

where l is the length of the microbend fiber, and $0 \leq q \leq 1$. A fiber with a perfectly absorbent jacket would have $q = 1$ and a sensitivity that depended linearly upon the sensing length. Silicone possesses a value of $q \approx 0.2$, and for black paint, $q \approx 1$.

In our experiments, special manufactured *pcs* glass optical fiber are used, whose core radius is 100 μm and clad radius 150 μm . The polymer coat also acts as the clad which is easy to be wiped off using chemistry method. In this experiment, optical fiber with a short segment coat being wiped off is embedded into the prepregs before cure. The bare segment surrounded by resin is just like black painted so as to have, which is mean, high microbending sensitivity during prepreg curing process.

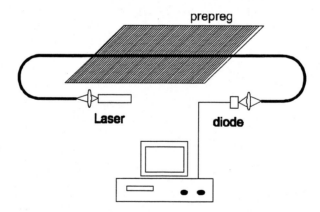

Fig. 1 Intrinsic fiber optic microbend sensor

3. DISCUSSION ON THE RELATIONSHIP BETWEEN CURE PROCESS AND MECHANIC PERFORMANCE

In the autoclave curing of laminates, the viscosity of the resin is very important factor to control the fabrication of composite structures. In the early stage, the viscosity of resin goes down to the nadir with temperature rising. To assure proper volume ratio of resin and fiber as well as the uniform distribution of the inner pressure, it is demanded to add pressure after the nadir of viscosity and before the gel point to ensure the quality of the products[1]. Besides, keeping the viscosity at nadir level is key to the voids transportation from inner to outsides of the prepreg, the longer that time maintains, the more voids will transport out under the differential pressure caused by vacuum. To keep the viscosity at nadir level as long as possible can be realized by holding the temperature invariable for a period of time. Of course, the viscosity will rise inevitably at last, but it does help to lower down the void ratio in the ultimate products. Meantime, by measuring change of viscosity, the end of cure can be confirmed which avoid partial cure or over cure. Thus viscosity sensing is the key to improve product quality and decrease costing in the manufacture process. It is proposed to measuring changes of viscosity of composite resin using fiber optic microbend attenuation in this paper.

4. COMPUTER SIMULATION

To find out the relationship between process mechanics and the ultimate products mechanic performance, a computer simulation program is developed based on process mechanics[1]. It is well know that the cure process is very complicate, according to former scientists's work it can be modeled as thermal—chemical principle、the resin flow and fiber function principle、the void action principle and the residual stress principle. The ultimate products quality depends on low void ratio, proper fiber-resin volume ratio, and low residue stress. Lowering down the residue stress is not in this presentation's consideration, we are trying to control the process to realize low void ratio and proper fiber-resin volume ratio. These two factors are mainly influenced by temperature and pressure routine, which is set up according to the resin viscosity evolution during the cure process, therefore to sense the viscosity of resin is critical to the cure process optimization.

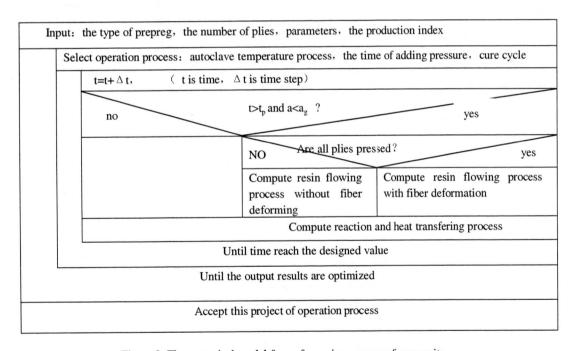

Figure 2. The numerical model frame for curing process of composite

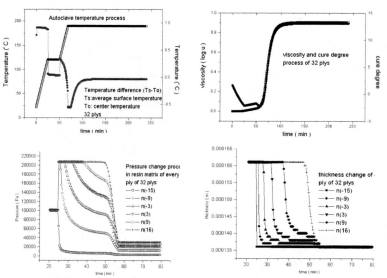

Fig.3 Output of the numerical model for 32 plies of prepreg cured in autoclave

5. EXPERIMENT

By the computer simulation we can determine how to adjust the process parameter as the viscosity changes. Once the viscosity during the cure process, the optimization of the process parameter becomes possible. With the new fiber optic sensing technology, the viscosity sensing is no longer complex and expensive. The viscosity of resin during cure can be revealed by microbend attenuation of the new fiber optic sensor. Measuring results of fiber optic microbend attenuation are given in Fig. 4 for carbon fiber bismaleimide(BMI) matrix QY8911-1 prepreg curing process. The test result of dielectric cure monitoring using the same material is given below as contrast. The lay up of the specimens with composite fiber orientation in contact with the optical fiber is: $[90_6/0(\underline{FO})/90_6]_S$. Curing press is 0.4mpa

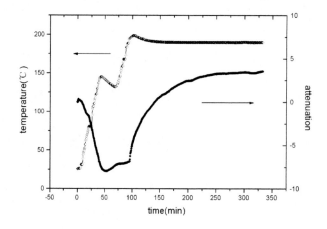

Fig.4 Test result of fiber optic sensor cure monitoring

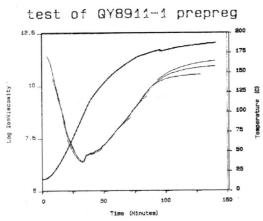

Fig.5 Test result of dielectric cure monitoring

Based on Gutowski model about resin flow-fiber elastic action[4,5], the prepreg laminate is considered as the elastic medium of nonlinearity with micro-holes, the added pressure is loaded by both resin and fiber. In the early stage, the load is supported only by resin matrix for the reason that the fiber does not distort. The resin can flow from the border of laminate under the difference of pressure between the center and the border of the laminate. The attenuation caused by microbend of fiber optic can be weaken; the prepreg plies will be compacted when the viscosity goes down to the lowest point, thus the fiber will be deformed and the elastic action will be caused. Part of loads are supported by fiber, the attenuation of output light is reinforced with the growth of microbend of fiber optic, while the average pressure in the resin goes down. The ratio of the load supported by fiber will grow little by little with the compaction of laminate, while the microbend of fiber optic will be reinforced little by little. The attenuation attains to the strongest and not varies any more when the cure process end. From the test's results it is proved that the fiber optic sensor that we developed is able to reveal the cure process of the advanced composite. As long as we can acquire the information inside the prepreg during the cure process, measures can be taken to control the process going as we wish.

6. EXPERT SYSTEM

In order to control the cure process, an expert system is constructed, its rules are set up on the basis of computer simulation results and experience. Now, a closed loop control system is finished. When the cure process starts, the expert system use the sensing information to control the process parameter, such as judgement the state of the resin, temperature rising speed, when to keep the temperature unchanged, when to add the pressure and how much pressure to be added, to meet our requirements.

7.CONCLUSION

A new type of fiber optic sensor based on optic fiber microbend attenuation is developed to sense the advanced fiber reinforced composite cure process. It is cheap, robust and easy to handle. A computer simulation program is developed to simulate the cure process in order to find out the influence of the process parameter on the ultimate products mechanic performance and to provide rules for the expert system, which is constructed to control the cure process. These three parts compose of a closed loop control system to aid the composite production. This system is proved very useful to the composite production since that has changed the cure process from an empirical approach to a scientific approach. It is sure that it will greatly decrease the cost, better the products quality and improve the efficiency of the composite production.

REFERENCE

1   Loos, A. C. and G. S. Springer, "Curing of Epoxy Matrix composites, Journal of Composite Materials", Vol, 17(1983), pp, 135—169.

2   Tim Clark, Herb Smith. "Microbend Fiber Optic Sensors", (Eric Udd ed.) John Wiley & Sons,Inc.,1994: 319~360

3   Suopajarvi P. et. al. "Indicating Cure and Stress in Composite Containers Using Optical Fibers", Optical Engineering. 1995, 34（9）: 2587~2591

4   Gutowski, T. G, "A Resin Flow/Fiber Deformation Model for Composites." SAMPE Quarterly, 16(4)(July 1985).

5   Gutowski, T G, Morigaki T, and Cai Z. "The Consolidation of Laminate Composites", Journal of Composite Materials. 1987, 21: 172~188.

Micromachined, SiC fiber optic pressure sensors for high-temperature aerospace applications

Wade Pulliam[*a], Patrick Russler[a], Richard Mlcak[b], Kent Murphy[a], Carrie Kozikowski[a]

[a]Luna Innovations, Blacksburg, VA
[b]Boston MicroSystems, Woburn, MA

ABSTRACT

This paper presents results of a micromachined, SiC/sapphire pressure sensor designed for propulsive environments. The completed sensor is 3mm square with a sapphire fiber through the back of the sensor. Included are results from a high-speed fiber optic signal processing system combined with sapphire fiber for use in fluctuating, high temperature environments. The sensor is designed to be capable of operating at the extreme temperatures and pressures of the next generation engines including ramjet/scramjets. These conditions far exceed the capabilities of conventional metal and electronic sensors. Fiber optic sensors offer the ability to increase the temperature range of these devices by removing the electronics of conventional sensors from the hot zone. Unfortunately, these conditions also exceed the capabilities of silicon and silica optical fiber. In contrast, silicon carbide has excellent mechanical, thermal and chemical properties for use in such environments, while the high operating temperature and optical quality of sapphire fibers and the inherent immunity of optical fiber sensors to electromagnetic interference make their use of particularly advantageous. Sensors made from a combination of these materials would be able to operate in almost any propulsive environment and allow valuable insight into flow regimes where little previous data is available.

Keywords: Pressure, silicon carbide, sapphire, high-temperature, micromachined, fiber optic

1. INTRODUCTION

Optical fiber systems have been developed during the past twenty-five years for primary applications in long-distance, high-speed digital information communication. Sensors using optical fiber have been developed over the past fifteen years for applications in the characterization of aerospace and hydrospace materials and structures, civil structures, and industrial process control and biomedical systems.[i,ii] Optical fibers are used as the field-sensitive elements in sensors for the measurement of environmental parameters such as strain, temperature, vibration, chemical concentrations, and electromagnetic fields. Their advantages for such measurements include 1) an inherent immunity to electromagnetic interference, 2) avoidance of ground loops, 3) the capability of responding to a wide variety of measurands, 4) excellent resolution compared to conventional foil strain gages, 5) the avoidance of sparks, which is especially important for applications within explosive environments, and 6) operation at temperatures of approximately 1073 K for silica waveguides and above 2173 K for sapphire waveguides.

Because these sensors are tolerant to extreme temperature, EMI, shock and vibration, and offer reduced weight and increase accuracy over conventional instrumentation, they have begun to replace conventional electrical sensors in many applications, and have been proposed for use in health monitoring aboard high performance aircraft. For example, the conditions propulsive environments, especially those of ramjets/scramjets, must be able to withstand pressures in excess of 50 atm., temperatures in excess of 2700° C, and heat flux levels as high as 5000 kJ/m²/s. These conditions far exceed the capabilities of conventional metal and electronic sensors.

Fiber optic sensors offer the ability to increase the temperature range of these devices by removing the electronics of conventional sensors from the hot zone. Unfortunately, such harsh conditions also exceed the capabilities of silicon, the common structural material in most micromachined sensors, which loses much of its strength by 550° C (1020° F) and is subject to severe oxidation rates at these high temperatures, and silica optical fiber, which softens at 900°C (1650°F). In contrast, silicon carbide (SiC) has excellent mechanical, thermal and chemical properties making it a material of choice for operation in such environments. SiC has extremely high melting (2760° C, 5000° F) and decomposition temperatures, has approximately three times the thermal conductivity, yield strength, hardness, and Young's modulus of silicon, and has exceptional chemical stability, with appreciable dissolution only in molten metals and molten metal salts. The 2050° C

---

* Correspondence: Email: pulliamw@lunainnovations.com; Telephone: (540) 953-4290; Fax: (540) 951-0760

(3720° F) melting temperature of optical quality sapphire fibers and the inherent immunity of optical fiber sensors to electromagnetic interference and radiation make the use of sapphire fiber-based sensors particularly advantageous. Sensors made from a combination of these materials would be able to operate in almost any flight or propulsive environment. The measurements provided by these sensors would allow valuable insight into flow regimes where little previous data is available.

2. EFPI-BASED SENSING METHODS

The transducing mechanism used by these sensors is based on the extrinsic Fabry-Perot interferometer (EFPI)[i] developed by Kent Murphy et al. EFPI-based sensors use a distance measurement technique based on the formation of a low-finesse Fabry-Perot cavity between the polished end face of a fiber and a reflective surface, shown schematically in Figure 1. Light is passed through the fiber, where a portion of the light is reflected off the fiber/air interface (R1). The remaining light propagates through the air gap between the fiber and the reflective surface and is reflected back into the fiber (R2). These two light waves interfere constructively or destructively based on the path length difference traversed by each. In other words, the interaction between the two light waves in the Fabry-Perot cavity is *modulated* by the path length. The resulting light signal then travels back through the fiber to a detector where the signal is *demodulated* to produce a distance measurement. Several different demodulation methods exist to convert the return signal into a distance measurement.

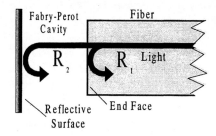

Figure 1. Extrinsic Fabry-Perot Interferometer Concept

Demodulation using single wavelength interrogation

A basic demodulation system using single wavelength interrogation is shown in Figure 2.

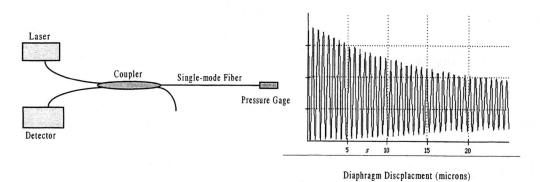

Figure 2. Basic Demodulation System and Typical Output of EFPI Pressure Sensor

A laser diode supplies coherent light to the sensor head and the reflected light is detected at the second leg of the optical fiber coupler. The output can then be approximated as a low-finesse Fabry-Perot cavity in which the intensity at the detector is,

$$I_r = \left| A_1 + A_2 \right|^2 = A_1^2 + A_2^2 + 2A_1 A_2 \cos \Delta\phi, \qquad (1)$$

if A_1 and A_2 are the amplitudes of R1 and R2, and $\Delta\phi$ is the phase difference between them. The output is then a sinusoid with a peak to peak amplitude and offset that depends on the relative intensities of A_1 and A_2. A phase change of 360 degrees in the sensing reflection corresponds to one fringe period. If a source wavelength of 1.3 μm is used, the change in gap for one fringe period is 0.65 μm. The drop in detector intensity is due to the decrease in coupled power from the sensing reflection as it travels farther away from the single-mode input/output fiber. By tracking the output signal, minute displacements are determined. The disadvantage of this type of demodulation system is the non-linear transfer function and directional ambiguity of the sinusoidal output.

An approach to solving these problems is to design the sensor head so that at the maximum gap the signal does not exceed the linear region of the transfer function. Confining the operation to the linear region places difficult manufacturing constraints on the sensor head by requiring the initial gap to be positioned at the Q-point of the transfer function curve. In addition to the difficult manufacturing constraints, the resolution and accuracy are limited when the signal output is confined to the linear region. To solve the non-linear transfer function and directional ambiguity problems, alternative signal demodulation approaches can be used such as 1) dual wavelength and 2) broadband interferometry using spectrometer based signal acquisition.[iii]

Demodulation using dual wavelength interrogation

One approach to the high-frequency demodulation system is based on dual-wavelength (2-λ) interrogation, and is suitable for single point or multiplexed configurations at frequencies up to 10 kHz and above. A working prototype of an architecture of this design is shown in Figure 3. The design utilizes advanced digital signal processing and dual optical channels to provide unambiguous, accurate, high frequency measurements. Two lasers of appropriate output wavelength are selected to generate quadrature (90° phase difference) signals for a given sensor air gap. The reflected laser signals from the sensor head are then separated out at the detector end using optical filters.

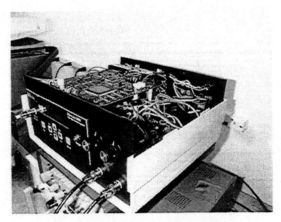

Figure 3. Prototype Demodulation System

The ambiguity in the interferometric intensity based demodulations systems can be removed with this quadrature output. [iv,v] Figure 4 shows a theoretical plot of the output signal that can be obtained with this system. As seen from the graph, when one signal is at a peak or valley, the other signal is in the linear region. In this manner, one signal always has a linear response. By monitoring the phase lead/lag relationship between the signals, the direction of gap movement is unambiguously determined. An output showing the phase lead/lag relationship at a direction change is shown in Figure 5.

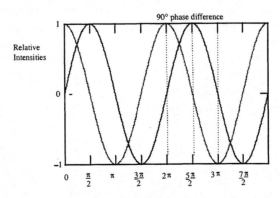

Figure 4. Theoretical outputs of quadrature signals obtained from a dual wavelength system.

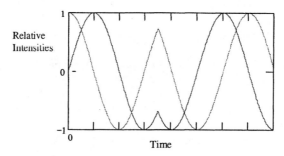

Figure 5. Two-λ system theoretical output showing the phase lead/lag relationship at a direction change.

In the region of 90° phase difference, gap values and directional changes can be monitored by plotting the two signals on opposite axes to create a Lissajous figure, as shown in Figure 6. By choosing source wavelengths a few nanometers apart, the range of unambiguous phase detection increases. By tracking the rotations around the circle and converting α to a displacement value, one finds the perturbation in terms of gap change. This relative displacement is then converted into a proportional voltage.

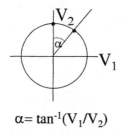

$$\alpha = \tan^{-1}(V_1/V_2)$$

Figure 6. Idealized Lissajous figure of dual-wavelength output.

This system is an improvement over the single wavelength system in that unambiguity is maintained over multiple fringes even if rapid changes in sensor output occur. However, both systems are limited to measuring only relative changes in the Fabry-Perot gap length. The absolute gap length is lost if the system loses power or is powered on at any condition other than the initial starting condition of the sensor.

Demodulation using spectral interrogation
Broadband spectral interrogation is an optical cross-correlation technique capable of determining very accurately the absolute path imbalance between two arms of an interferometer. For the case of an EFPI-based sensor, white-light interferometric techniques provide the exact gap separation, D, between the fiber end face and a reflective surface forming the Fabry-Perot cavity. In more detail, a broadband, white light source, as depicted in the upper left of Figure 8, is transmitted to the sensor, where it becomes modulated by the Fabry-Perot cavity. The modulated spectra, as shown in the upper right of Figure 8, are then physically split into its component wavelengths by a diffraction grating, which are measured by a charged-coupled device (CCD) array, as shown in Figure 7. The spectra are then off-loaded from the CCD array, and an optical path length is calculated from the spectra using the Luna algorithm. The main part of the algorithm is an FFT, which transforms the signal from a wavelength domain to a gap domain. The output of this is a plot that has numerous peaks, as shown in the bottom of Figure 8. The location of the maximum of the main peak, found with a peak search, is the absolute optical gap of the EFPI cavity.

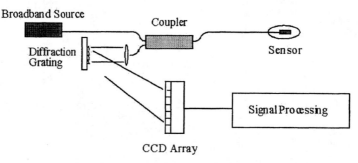

Figure 7. Absolute Fiber Optic Demoduation System.

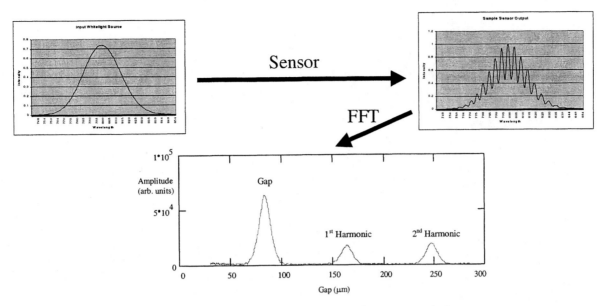

Figure 8. Depiction of spectral interrogation system method.

The determination of an absolute gap is extremely useful. Although most transducers are of a relative nature to a nominal value, the absolute gap removes the ambiguity seen with the previous system. Also, the system can be turned off, and back on, and the data can be gathered again from that point, without having to determine the nominal point. Because the system is gathering data from a range of wavelengths, and not just one, the resolution (estimated to be 200 picometers) and accuracy of the system are robust where other systems are prone to error, such as micro-bends in the fiber and noise/fluctuations of the source.

This system provides absolute output, and the accuracy and robustness of this system allow its use in calibration and characterization effort of new sensor types. However, the computational time necessary for the algorithm of this system makes use of this optical signal processing system difficult for the intended environment of the SiC pressure sensor. The motivation of this development is for high temperature/high EMI environments, ones in which conventional sensors can not operating. If one was only interested in low frequency measurements, then a conventional pressure sensor could be removed from the environment by a pressure line. Therefore, these SiC sensors must operate at high frequencies and yet be characterized accurately, requiring that the sensors operate well with both the dual-wavelength and broadband spectral interrogation systems. In fact, since the design temperature of the SiC pressure sensor is above the long-term operating temperature of silica, both of these systems must be operated through sapphire waveguides.

3. SAPPHIRE FIBER OPTICAL SIGNAL PROCESSING

Sapphire has a very high melting point (2050° C, 3720° F) with excellent optical qualities, and is therefore a critical component in the development of a high-temperature optical sensor. Inexpensive silica fiber will be used in the cool zones below 700° C and sapphire fiber will be used over the short distances to survive the high temperatures up to and including the sensor head. As mentioned above, it is important for the sapphire demodulation to operate with both the broadband spectral interrogation and the dual-wavelength systems, and each has been demonstrated to work effectively at Luna for this project.

Dual-Wavelength Demodulation

A single-mode silica fiber was used to inject light from the 2-λ system into a 180 μm diameter sapphire fiber. The other end of the sapphire fiber was inserted into an existing Luna fiber optic pressure sensor that had been fitted with a SiC diaphragm. To demonstrate a sapphire optical fiber demodulation system on a SiC membrane, this pressure sensor was mounted on a translation stage to control the gap length between the end of the sapphire fiber and the inside surface of the diaphragm. The gap was varied to produce modulated return signals from the two light sources. Figure 9 shows the signals are in quadrature, and also shows a "turn-around" point where the gap movement changed direction. Deviations of the detected signal from the actual displacement are due to error introduced from manual adjustment of the micropositioner.

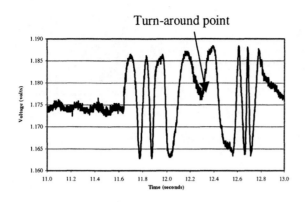

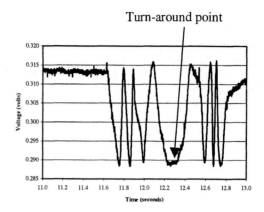

Figure 9. Quadrature phase-shifted signals obtained with simulated pressure sensor with SiC diaphragm and sapphire fiber.

Absolute Spectral Demodulation

Even though the dual-wavelength system has been demonstrated with an elementary coupling between the silica and sapphire fibers (necessitate only careful alignment), the system relies on only two wavelengths. The broadband spectral system requires the discrimination of a large number of wavelengths, whose individual power content needs to remain constant relative to each other. The sapphire waveguides have no cladding, which means that they cannot be bent significantly without disrupting the optical path. The lack of cladding also makes the sapphire waveguide effectively larger in diameter than the silica fiber core, so that the diameters of the optical paths in both waveguides do not match. Without a method of compensating for this difference in diameters, much of the light reflected down the sapphire waveguide from the EFPI sensor does not couple back into the small diameter silica fiber. Successful silica/sapphire fiber coupling was the key to acquiring optical measurements through a sapphire waveguide.

A coupling method was developed during the program using a graded index (GRIN) lens fixed between the ends of a 2.5 cm long, 180 μm diameter sapphire waveguide and a commercially available optical fiber, as depicted in Figure 10. This experimental arrangement is shown schematically in Figure 11. Light was passed through the coupler from the silica fiber into the sapphire fiber. Reflected light followed the reverse path. A Fabry-Perot cavity was formed between the end of the sapphire waveguide and a reflective surface. This produced a return spectrum (shown in Figure 11) of sufficient quality to determine the cavity length using the broadband spectral demodulation system. This was the first successful demonstration of the broadband spectral processing system using sapphire as a waveguide.

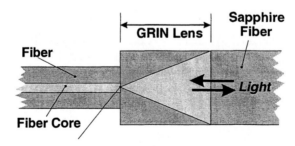

Figure 10. Schematic representation of a 0.25 pitch

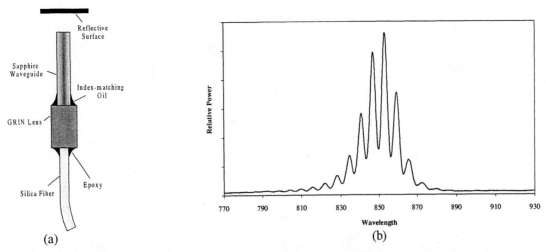

(a)   (b)

Figure 11. Silica/sapphire fiber coupler (a) and return spectrum through sapphire waveguide (b)

In order to verify the performance of the sapphire waveguide in a more demanding application, a current Luna Si pressure sensor, as describe in the next section, was constructed with a sapphire waveguide using a similar coupling method. This is shown schematically in Figure 12. The absolute demodulation system was used to calibrate this sensor using methods and equipment identical to that used to calibrate silica waveguide pressure sensors already in use. The resulting calibration is shown in Figure 12.

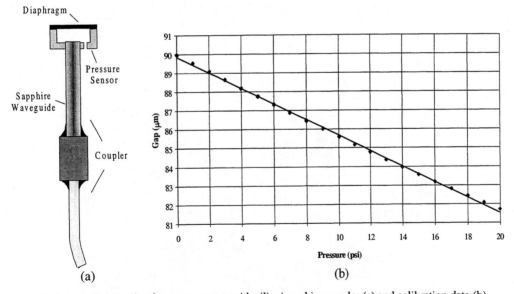

(a)   (b)

Figure 12. Schematic of pressure sensor with silica/sapphire coupler (a) and calibration data (b)

4. SILICON CARBIDE SENSOR DESIGN

The micromachined silicon carbide-sapphire fiber optic pressure sensor design is shown in Figure 13. The design is similar to a proven Luna pressure sensor design based upon a silicon membrane, glass body and silica fiber as illustrated in Figure 14. This design has proven to be very successful and is the basis of our current pressure sensor product line. The SiC sensor is 3X3 mm and 500 microns thick, providing a small envelope for detailed surface pressure measurements. Furthermore, these sensors can be easily packaged into arrays, as already demonstrated with Luna's silicon/glass pressure sensors[vi], for complete surface pressure mapping in propulsive and other extreme environments.

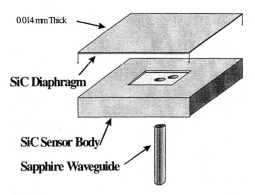

Figure 13. Sketch of the SiC pressure sensor to be developed by Luna and Boston MicroSystems.

Again, the basis of the fiber optic pressure sensors is shown in Figure 14. An optical fiber and glass tube fiber spacer are bonded in the center opening of the sensing element using epoxy. The optical gap between the bottom of the diaphragm and the end face of the fiber is a Fabry-Perot cavity. As described above, light interference resulting from the internal reflection of light at the fiber end face (R_1) and reflected light off of the bottom surface of the diaphragm (R_2) is used to monitor the optical path length. The optical gap varies with diaphragm deflection, which in turn varies with applied pressure. The reference port on the bottom of the sensing element acts as a vent through which air can pass to maintain a constant pressure on the reference side of the diaphragm. This reference port can be sealed and the interior evacuated to form an absolute sensor.

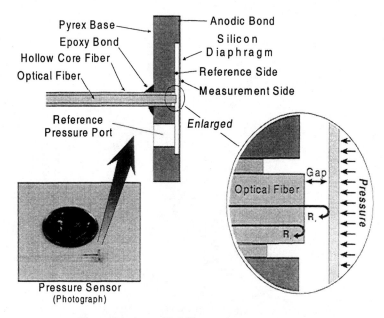

Figure 14. Assembled fiber optic pressure sensor.

Boston MicroSystems, Luna's partner in the development of the silicon carbide-sapphire fiber optic pressure sensor, has now delivered the first prototype silicon carbide membrane chips for the sensor. Micromachining of the SiC sensor body is accomplished by ultrasonic machining in the same manner as used to produce the glass body for Luna's commercially-available pressure sensors. The SiC diaphragm is then deposited on the body through a proprietary process. While most of the fabrication issues have now been addressed, work continues to optimize the deposition to achieve minimal membrane stress for linear sensor response.

5. SILICON CARBIDE MICROMACHINED PRESSURE SENSOR RESULTS

In order to demonstrate stability of the SiC membrane chips at elevated temperatures characteristic of propulsive environments, one of the devices was placed in a kiln and heated to 1000* C for one hour. As expected, neither the membrane, chip body, or membrane body interface was affected by the high temperature exposure, demonstrating the ability to survive in the intended environment. The SiC chips were then used to fabricate fiber optic pressure sensors using standard silica fibers, a shown in Figure 15. Sensors were made with both the reference pressure port sealed to form absolute pressure sensors, and with the reference port open to form relative pressure sensors.

Luna's broadband optical signal processing system was used to measure the spectrum return of these sensors, shown in Figure 16a. The return spectra differ significantly from those of silicon membranes due to the optical transmissivity of SiC. The signal seen is actually multiple optical gap superimposed upon themselves, which can more easily be seen in the Fourier transform of the signal, Figure 16b. The first peak is the internal reflection within the diaphragm (25 mm), the second is the air gap that is of interest (~68 mm), and the third is the combination of the first two (93 mm). Beyond those peaks are the harmonics of the first three peaks. This presents some difficulty for the algorithm, as the diaphragm peak is sometimes the strongest and the first and second harmonic actually bracket the air gap peak of interest. Algorithm improvements and diaphragm reflection abatement are being pursued currently to rectify these difficulties.

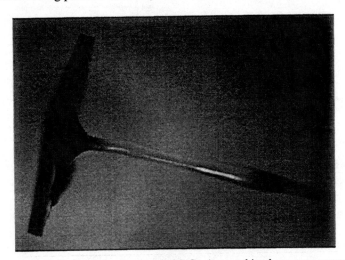

Figure 15. Photograph of assembled SiC micromachined pressure sensor.

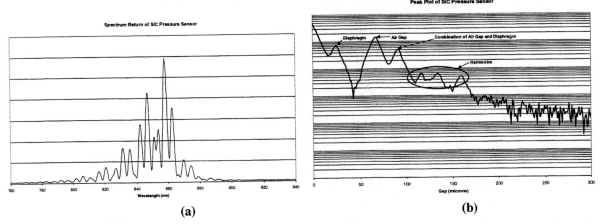

(a)   (b)

Figure 16. Optical signal from the SiC pressure sensor, a) spectrum of signal from broadband optical source through the silica fiber, b) FFT of that spectrum showing the diaphragm and air gap, as well as the combinations and harmonics.

The prototype absolute and relative SiC pressure sensors were then calibrated, with the results presented in Figure 17a and b. Both sensors follow a cubic trend as is predicted by thin wall theory, however the concavity of the calibration curve was opposite of that seen in the Si pressure sensor. This is attributed to the internal stress in the membrane that results in an outward bulging of diaphragm in its steady-state position. In the stress-free Si sensors, increasing pressure causes the diaphragm to stretch more, decreasing the deflection/psi as the magnitude of the pressure increased. However, with the outwardly bulging diaphragm, it would be stretching less, increasing the deflection/psi as the magnitude of the pressure increased. At sufficiently high pressures, such outward bulging membranes may tend "pop" into being inwardly bulging, and is hypothesized to be a failure mechanism for the existing devices. Boston MicroSystems is now preparing the next generation micromachined SiC membrane chips, using a newly developed method providing greatly reduced internal stress in the SiC diaphragm to eliminate these problems.

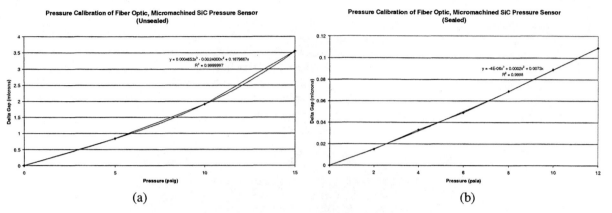

(a)   (b)

Figure 17. Pressure calibration of the fiber optic, micromachined SiC pressure sensor, a) calibration of a relative gage where the reference pressure port was open to atmospheric pressure, and b) one of an absolute gage where the port was sealed.

6. CONCLUSIONS

A micromachined SiC, fiber optic pressure sensor has been developed for use in the extreme temperatures and pressures of propulsive environments. Since silicon fibers cannot perform at these temperatures, optical signal processing using sapphire waveguides were developed for this application. The use of two types of optical signal processing systems with sapphire fibers has been demonstrated, one for accurate measurements used in sensor characterization, and another used for high-frequency measurements necessary in such propulsion testing. The combination of the sapphire waveguide and SiC membrane chip provides for a fiber optic pressure sensor capable of operating above 1100° C.

ACKNOWLEDGEMENTS

The authors would like to acknowledge support for this work under Air Force contract out of AFRL (Wright-Patterson), Propulsion and Power Directorate. Finally, the authors would also like to thank Dr. Joseph Schetz of Virginia Tech for his contributions to the development.

REFERENCES

[i] Murphy, K., Gunther, M., Vengsarkar, A., and Claus, R. O., "Quadrature phase-shifted, extrinsic Fabry-Perot optical fiber sensors," *Optics Letters*, Vol. 16, No. 4, pp. 173-275, February 1991

[ii] Claus, R.O., Gunther, M.F., Wang, A., Murphy, K.A., "Extrinsic Fabry-Perot Sensor for Strain and Crack Opening Displacement Measurements form −200 to 900 C", Smart Material Structures, Vol. 1, pp. 237-242, 1992

[iii] Dakin, J., and Culshaw, B., Optical Fiber Sensors: Principles and Components, Artech House, Boston, MA, 1988.

[iv] Ezbiri, A., Tatam, R., "Passive Signal Processing of Miniature Fabry-Perot Interferometric Sensors Using a Dual-Wavelength Source", 10th Optical fibre Sensors Conference, pp. 335-338, 1994

[v] Gelmini, E., Minoni, U., Docchio, F., "Tunable, Double-Wavelength Heterodyne Detection Interferometer for absolute-Distance Measurement", Optics Letters, vol. 19, no. 3, pp. 213-215, 1994

[vi] Borinski, J., Meller, S., Pulliam, W., Murphy, K., "Optical Flight Sensors for In-Flight Health Monitoring", SPIE Proceedings, vol. 3986, Newport Beach, CA, Mar. 6, 2000

Seismic fiber optic multiplexed sensors for exploration and reservoir management

Mark H. Houston

Litton Guidance & Control Systems,
a division of Litton, Inc.

ABSTRACT

Reliable downhole communications, control and sensor networks will dramatically improve oil reservoir management practices and will enable the construction of intelligent or smart-well completions. Fiber optic technology will play a key role in the implementation of these communication, control and sensing systems because of inherent advantages of power, weight and reliability over more conventional electronic-based systems.

Field test data, acquired using an array of fiber optic seismic hydrophones within a steam-flood, heavy oil-production field, showed a significant improvement (10X in this specific case) in subsurface resolution as compared to conventional surface seismic acquisition. These results demonstrate the viability of using multiplexed fiber optic sensors for exploration and reservoir management in 3-D vertical seismic profiling (VSP) surveys and in permanent sensor arrays for 4-D surveys.

Keywords: Fiber optic, sensors, multiplexed, VSP, 4-D, Mach-Zehnder, interferometer, 3-D VSP.

1. INTRODUCTION

In the 90's the introduction of new imaging and drilling technologies into the Exploration & Production industry significantly mitigated the upward pressure on oil commodity prices by decreasing finding and production costs. For example, Figure 1 shows a plot of exploration-well completion rates (%), crude oil prices ($/BBL) and the average worldwide rig count for the years 1975 through 1999.[1.]

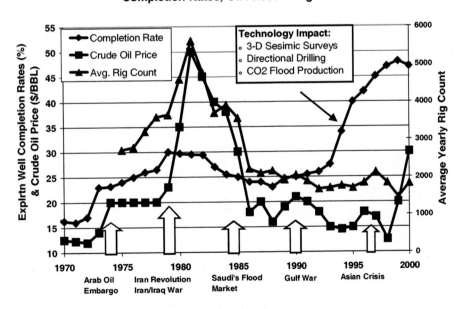

Figure 1. Completion rates, oil prices and rig counts over the last 30 years reflect the influence of international events. Technology has served to mitigate the long-term price rise.

1.1 Inflation Adjusted Oil Prices – The Industry Driver

Oil prices are the driver and the barometer of the oil industry. When prices fall, look for stormy times ahead; when prices abruptly rise, expect exploration and production activity to head upward. Oil is a commodity whose market price gyrations are often triggered by international and political conflict. Notwithstanding these periods of price spikes, technology has served to moderate the long-term rise in oil and gas prices.

1.2 Average Rig Count – A Measure of Industry Activity
The rotary rig count is the average number of drilling rigs actively exploring for oil and gas, thus the rig count is a measure of the activity and confidence of the upstream segment of the oil and gas industry. The worldwide rig count closely tracks the crude oil price.

1.3 Exploration Well Completion Rate – A Measure of Drilling Success Rate
After an exploratory well is drilled, the well is classified as a dry or an oil or gas well, and only wells that are judged economic are completed. The percentage of wells completed as oil or gas wells is frequently used as a measure of success. The rise in the completion (success) rate in the 1970's can be attributed to activity generated by rising prices. The increases in the 1990's, however, are clearly the result of improved exploration technology such as 3-D seismic, directional and horizontal drilling and CO_2-flood production.

1.4 Active Reservoir Management – Success for the Millenium
In the 00's the introduction of active oil reservoir management practices will be required to effect a similarly favorable financial impact on oil prices. To a large extent, the success of these improved reservoir management systems will depend on reliable downhole communication, control and sensor networks. Fiber optic technology will play a key role in the implementation of these communication, control and sensing systems because of inherent advantages of power, weight and reliability over more conventional electronic-based systems.

1.5 Fiber Optic Sensors to Image Additional Oil-In-Place
To date, about a third of the US oil reserves have been produced with another 9 percent extractable by conventional or simple EOR methods.[2] As illustrated in Figure 2, the remaining oil distribution consists of mobile oil in unswept zones (16%) and of immobile oil, which may be partially recoverable by Enhanced Oil Recovery (EOR) methods.[3] Old field redevelopment targets this unswept mobile oil for reserve growth by infill wells, side-tracks or recompletions. New field development aims to avoid unswept mobile zones through optimum placement of wells and through active fluids management. In both cases the reservoir engineer requires a detailed characterization of the reservoir with a spatial resolution that is unattainable using surface seismic imaging techniques. Downhole sensor systems are required to better image the reservoir structure, rock properties and fluids distribution.[4]

Downhole fiber optic sensor systems are an enabling technology that will allow the reservoir engineer to better image and characterize the reservoir. The capabilities and reliability of fiber optic sensors allow volumetric sampling of reservoir characteristics with a spatial resolution and a time-span unattainable using currently available conventional technologies.

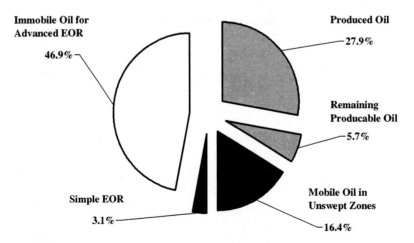

Figure 2. US oil distribution for targeted drilling prospects. Unswept oil recoverable by
conventional methods represents billions of dollars in additional in-place reserves.

2. FIBER OPTIC ACOUSTIC SENSORS

Fiber optic sensors to measure temperature and pressure have been previously used in well-bores, but acoustic sensors for downhole imaging use have proven to be more difficult to build, principally because of the large dynamic range needed for seismic imaging. The dynamic range required for temperature sensors is often quoted as 60 to 70 dB, whereas, seismic sensors require a dynamic range of 120 to 130 dB.

2.1 Hydrophone Mach-Zehnder Configuration

One acoustic sensor that exhibits the necessary dynamic range for seismic applications is the Mach-Zehnder hydrophone interferometer as shown in Figure 3. In this type of sensor, coherent light enters the sensor and is split into two optical paths. One signal path passes through an acoustically isolated reference leg and the second signal path passes through optical fiber wound around an acoustically sensitive mandrel. Minute variations in the ambient pressure change the diameter of the mandrel and due to stretching of the wound fiber, also change the length of the sensor leg optical path. Reference and acoustic leg signals are recombined at the output of the sensor. The change in path mismatch between the two sensor legs results in an interference signal or phase change at the output of the sensor. The optical signal phase changes then map into acoustic sound levels.

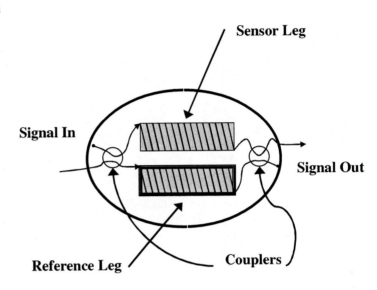

Figure 3. Mach-Zehnder Interferometer hydrophone configuration enables 130 dB dynamic range sensor.

The fundamental "output" of the fiber optic acoustic sensor is phase change, then the usual hydrophone sensitivity and recording system noise are expressed slightly differently from conventional piezo-electric systems. The fiber optic hydrophones have a design sensitivity in micro-Pascals per radian (μPa/rad.). The FDM electronics detect the phase change with a design demodulator scale factor in radians per volt (rad./V). The resulting output sound level is equivalent to the conventional units of micro-Pascals per volt ((μPa//V).

Fiber optic hydrophones have been designed and constructed to span a range of pressures, sensitivities and temperature conditions. For example, pressure compensated hydrophones will operate to 20,000 psi; the borehole hydrophone has been designed to record the minimum expected ambient noise levels for quiet well conditions at a temperature of 150 °C for an extended period of time.

2.2 Geophone – All Fiber Accelerometer

Prototype all-fiber accelerometers, which can be used to record 3-component seismic signals, have been constructed using the same Mach-Zehnder configuration so that they will operate with the same electronics as the fiber hydrophones, albeit a different scale factor. In this configuration, minute variations in acceleration of the sensor arm, change the length of the sensor arm path as compared to the reference arm. In exactly the same manner as the hydrophone interferometer, the fiber "geophone" sensor outputs a phase change with a design sensitivity in micro-g's per radian (μg/rad.).

2.3 Geophone – Electrical/Optical Hybrid

Figure 4 illustrates an alternative hybrid configuration for a passive fiber optic telemetry geophone sensor system. In this hybrid system, a standard geophone serves as the principle ground motion detector. A geophone uses the motion of a spring-supported coil in the field of a permanent magnet to generate an output voltage. In typical electronic seismic acquisition systems, the geophone signal is input into a preamplifier and the boosted signal is converted to a digital format using an analog-to-digital (A/D) converter. The digital signal is then transmitted to the receiver electronics where the signals are decoded and recorded. In the E/O Fiber Optic Hybrid seismic acquisition system, the geophone signal is input into a passive E/O converter and the geophone voltage generates a phase change in the sensor arm of a Mach-Zehnder interferometer. The topside fiber optic electronics then demodulate the optical signal changes and map the changes into the equivalent digital

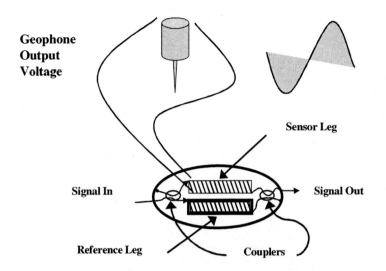

Geophone Output Voltage

geophone response. The hybrid "geophone" configuration outputs a phase change with a design sensitivity of micro-V/m/s per radian (μV/m/s/rad.).

Figure 4. The E/O passive converter transforms the geophone voltage output signal into an optical phase change. This hybrid configuration allows standard geophones to be used with the FDM electronics.

3. SENSOR/ELECTRONICS SYSTEM DESIGN

3.1 FDM Telemetry

The topside Frequency Domain Multiplexing (FDM) electronics, consisting of frequency synthesizer, laser source modules, optical receivers, and demultiplexing-demodulation modules, contain all elements required to support the sensors. Low-noise, source lasers are used to continuously interrogate multiple sensors over common fiber paths by using slightly different frequencies or "colors" for each sensor. The architecture of Frequency Domain Mulitplexing is shown in Figure 5. Modulator electronics control the carrier modulation of the lasers. The center frequency and carrier modulation are used to demultiplex and demodulate the returning signals from individual laser/sensor pairs.

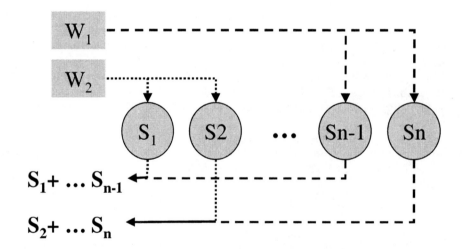

Figure 5. Frequency Domain Multiplexing can be used to multiplex multiple sensors on common fiber channels.

3.2 No Downhole Electronics

One advantage for the fiber optic acoustic system is that the remote sensors contain no electronics. All electronics are housed in the "topside" electronics cabinet, which is usually enclosed in a benign, humanized environment. Since simplicity or lack of complexity and reliability go hand-in-hand, the passive nature of the outboard part of the system exposed to the harsh environment is an immense advantage. While conventional electronics can be packaged to resist mechanical shock and corrosive fluids, elevated temperatures above about 125 °C provide a severe challenge to commercial-off-the-shelf (COTS) electronic components. Components do exist that extend this temperature range, but they are limited in selection and expensive in design. Current temperature limits on the fiber optic sensors are constrained by the packaging materials, not by the fiber; 200 °C sensors are not a stretch.

4. FIBER OPTIC BOREHOLE FIELD TEST

4.1 Field Test Set-Up

A field test of a prototype fiber optic borehole system was conducted at Texaco's Kern River field in Bakersfield, California. We recorded several cross-well data sets between two observation wells separated by a distance of 518 feet. Figure 6 shows

Figure 6. Fiber Optic Borehole system test equipment.

the components of the fiber optic borehole test system. The wireline truck contains the electronics and the 10,000-ft. lead-in cable that connects to the downhole array . This armored cable of 0.454-in. diameter has the same diameter as common wireline logging cable, and can be handled on standard wireline trucks.

Figure 7 shows the test geometry. A four-channel array with a hydrophone spacing of five feet was lowered into a 4"-cased

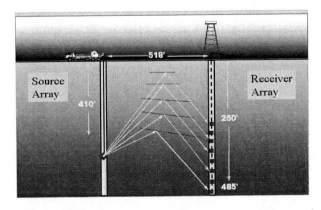

Figure 7. Test site geometry for common source gathers.

well. A 10,000-ft. lead-in cable connected the array to the FDM electronics mounted in a standard wireline truck. An Ethernet link transferred the demodulated seismic data in SEG-D format to a workstation where the raw data were recorded, displayed and processed on-site. The source was lowered into a 7"-cased observation well. The downhole source was a single-component axial vibrator with power source, cable, and control equipment mounted on a large trailer. This clamped source provides a large amount of seismic energy without harming the cement-casing interface.

4.2 Data Recording

A preliminary data set was recorded with a 10-s, 10 to 500-Hz upsweep in a common-receiver geometry. Figure 8 shows a composite, single-fold (100%) data record from one hydrophone channel after band-pass filtering to remove pump-jack noise below 30 Hz. Data were recorded with the receiver positioned at a depth of 400-ft. and the source depth varying from about 250 ft. to 720 ft. Source depth interval was 10 ft. The reflections underneath the first arrivals are not readily separable from noise in this view.

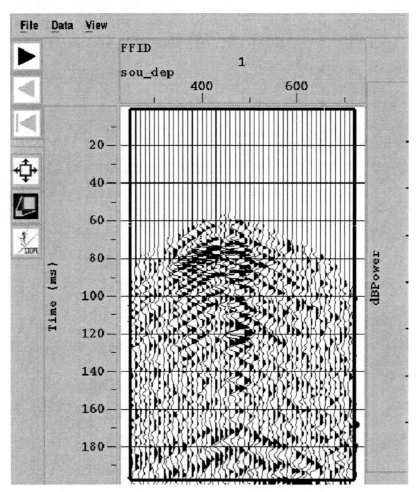

Figure 8. Data recorded as a common receiver gather. This single-fold (100%) record
was band-pass filtered to remove pump-jack noise below 30 Hz.

A second sequence of gathers was acquired as a common-shot fan with the source at 410 ft. and with receiver depths at 5-ft. intervals between 250 and 485 ft. Each seismic record in this data set is the correlated output of a 10-s, 10 to 500-Hz upsweep. Unfortunately, the data of this test geometry was limited because of high temperature steam leakage into the source observation well. Elevated temperatures limited the operating depth of the source to a maximum of 410 ft.

5. RESULTS AND DISCUSSION

5.1 Correlated Data and Spectra

Figure 9a (on the left) shows correlated data recorded with the source at 410 ft. and the source receiver depths ranging from 250 ft. to 485 ft. The horizontal scale represents receiver depth from the surface. The trace interval is 5 ft. The data clearly show strong reflectors beneath the first arrivals. The numerous events slanting away from the direct arrivals are reflections; the point of intersection with the direct arrival is the depth of the reflector at the borehole.

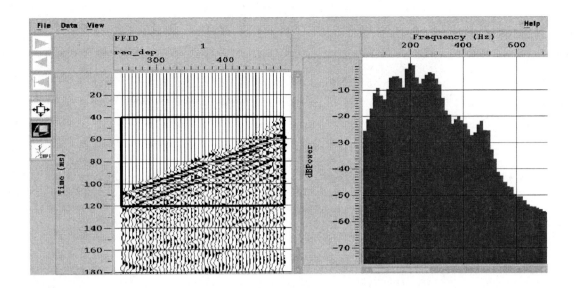

Figures 9a (left) and 9b (right). Common source gather recorded with a source depth of 410 ft. A top mute was applied and 30 Hz low-cut filter to remove pump jack noise. Reflections are very strong under the first arrivals. Amplitude spectrum is derived from data outlined in the box window. The source sweep was 10-500 Hz.

Figure 9b (on the right) shows the amplitude spectrum of data outlined within the box of Figure 9a. Peak energy of the data is 200 Hz with useful energy to about 500 Hz. Although attenuation and scattering of seismic waves are high in the Kern River Formation, the source strength and receiver sensitivity are sufficient to provide signal bandwidth and data resolution far superior to what has been achieved in this prospect with surface seismic techniques.

5.2 Stacked Migrated Data Image

Figure 10 shows a migrated depth image of the down-going reflections. The horizontal scale represents distance from the receiver well. The trace interval is 5 ft. The vertical scale has a depth resolution of about 5 ft.

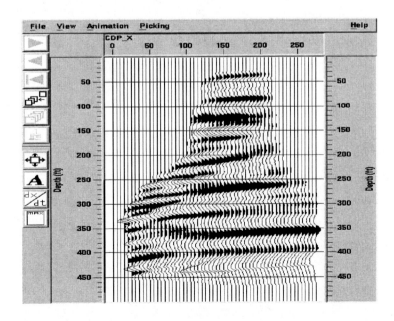

Figure 10. Depth image generated by migration of downgoing reflections in crosswell data with downhole source at 410 ft, and fiber optic receivers from 250-485 ft at 5-ft interval.

5.3 Comparison with Geologic Model

The migrated depth image was compared to a geologic cross-section derived from a standard suite of openhole well logs, surrounding temperature and saturation profiles and known stratigraphic characteristics of the Kern River formations. The correlation between the measured seismic data and the geostatistical derived data is very good over the vertical interval that was imaged, moreover, in the space between the wells, the seismic image shows continuity and pinch-out features that geostatistically-derived models cannot reveal.

6. CONCLUSIONS

Steam injection has enabled Kern River to become one of the most prolific oil producers in the continental US. However, the cost of the steam injection, combined with the relatively lower value of the heavy crude, has forced the asset manager to place an extra emphasis on production efficiency and cost control. Surface seismic imaging has not been an effective tool for Kern River, but these tests have demonstrated that a borehole hydrophone cross-well system together with a powerful downhole seismic source can provide useful high resolution images. Fiber optic sensors have inherent advantages of power, weight and reliability over more conventional electronic-based systems that will engender economic advantages for seismic 3-D VSP and cross-well surveys, but more importantly, for permanent sensor installations that will be used for 4-D surveys.

ACKNOWLEDGMENTS

We thank Texaco for access to their Kern River field and their on-site support of this test program. We thank Litton Guidance & Control Systems, Paulsson Geophysical Services, Inc. (P/GSI) and Texaco, Inc. for permission to publish the results. Thanks also to the LG&CS development team for their hard work and long hours in the field to complete the survey under strict time constraints, and to Brian Fuller of P/GSI who processed the seismic images.

REFERENCES

1. J. L. Williams, "History and analysis – crude oil prices", WTRG Economics, 2000.
2. A. Brooks, Herzog, L., and R. Ford, "Technology's role in the oil patch", Equity Research Paper, CIBC Oppenheimer, New York, N.Y., March 23, 1999.
3. K. J. Weber, "The prize – what's possible?", *Petroleum Geoscience*, **5**, 135-144, 1999.
4. E. W. Fisher, "Can the us oil and gas resource base support sustained production?", *Science*, **236**, pp. 1631-1636, 1987.

Distributed temperature sensing (DTS)
to characterize the performance of producing oil wells.

Glynn Williams[a], George Brown[a], William Hawthorne[a]
Arthur Hartog[b], Peter Waite[b].

[a]Sensa, Andover, Hampshire, United Kingdom

[b]York Sensors Ltd, Chandlers Ford, Hampshire, United Kingdom

ABSTRACT

This paper describes how distributed temperature sensing (DTS) based on Raman Scattering is being used as an in-situ logging technique in oil & gas wells. Traditional methods of gathering production data to characterize oil & gas well performance have relied on the introduction of electric logging tools into the well. This can be an expensive process in highly deviated or horizontal wells and usually results in the well being shut-in with the loss or deferment of hydrocarbon production. More recently permanently placed pressure sensors based on CMOS technology have been used, but these systems do not easily deliver distributed measurements and reliability has been found to be poor.

In order to apply DTS technology to this new type of application a number of new processes and methods have been developed. These include improved optical fiber coatings for high temperature applications, adaptation of fiber-blowing techniques techniques pioneered by the telecoms industry for installation and advanced interpretation techniques that allow distributed temperature profiles in the oil & gas well to give valuable information about flowing profiles, fluid properties and the condition of artificial lift equipment that is used to enhance production.

Keywords: Distributed Temperature Sensing (DTS), Raman Optical Time Domain Reflectometry (OTDR), optical coatings, Hydrogen darkening., Brillouin scattering. Extrinsic Fabry-Perot Interferometry (EFPI), oil and gas well logging.

1. INTRODUCTION

Research into the use of optical fibers as transducer elements can be traced back at least 20 years. It has given rise to a plethora of devices in the laboratory, but rather fewer have found practical commercial applications. The primary reasons for this lack of acceptance are the existence of alternative transducers that in many cases provide adequate solutions, together with the usually high cost of optical fiber sensors.

However, certain fiber-optic sensors are uniquely placed to address previously unsolved measurement problems. Examples include measurements in very high electro-magnetic fields, such as in power transformers and energy cables, where fiber-optic temperature and current sensors have found a niche owing to their immunity to EMI. The fiber-optic gyroscope has also proved to be a versatile device, capable of being designed either to meet stringent cost targets with performance adequate for in-car inertial navigation, or alternatively, to exceed the sensitivity, drift and scale factor stability of the best mechanical or laser gyroscopes. This paper addresses a class of sensors, namely, that of distributed sensors[1], which although unique to optical fibers presents similarities to techniques such as OTDR[2], Radar, Lidar or Ultrasonic reflectometry. Specifically, the operating principles and the applications of distributed temperature sensors (DTS) are discussed in the context of their applications in the upstream oil & gas industry.

The principal characteristic of distributed sensors is that they provide a complete measurand profile of the fiber that forms the sensor. Thus typically, a single instrument connected to an optical fiber several km long will provide thousands of temperature readings at points spaced as closely as 1m apart along the entire length of the fiber. In addition, the instrumentation interrogates fibers of the same designs as are used for telecommunications; these fibers are thus inexpensive and manufactured in large volumes to very high standards of optical transmission, dimensional accuracy and mechanical strength.

1.1.   Principles of operation of distributed temperature sensing

Most commercially available distributed temperature sensors are based on the principles of Raman Optical Time-Domain Reflectometry (OTDR) [3,4] in which a short pulse light is launched into the sensing fiber. From the backscattered light spectrum, a very weak, but temperature-sensitive band (the so-called anti-Stokes Raman band) is selected. The time dependence of the latter's intensity is a direct measure of the fiber temperature at a position determined by the round-trip propagation time. The efficiency of the process used, spontaneous Raman scattering, is independent of the probe pulse intensity and this simplifies considerably the signal processing required, compared with sensors using non-linear effects, such as stimulated processes5 or the Kerr-effect[6]. The basic optical arrangement is illustrated in Fig. 1.

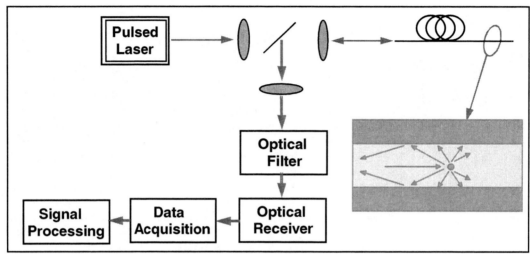

Figure 1: Basic arrangement of a Raman Distributed Temperature Sensor

In practice, a number of normalization steps are required to eliminate the effects on the measured value, of losses in the fiber, and of variation in fiber characteristics. Thus in practical systems a second wavelength (the Stokes Raman band) is frequently used to provide a reference. In multimode systems, a double-end measurement, where the data is collected successively from each end of the sensing fiber (double-ended operation), provides a supplementary referencing method and the combination of the dual wavelength and double-ended measurements provides a very robust determination of the temperature, which has proved to be resilient to changes in fiber or connector losses.

The criteria by which a distributed sensor is measured include its spatial resolution, i.e. its ability to distinguish two closely spaced events along the fiber. Also of interest is the range, i.e. the length of fiber that can be addressed by a single instrument; together, the spatial resolution and the range determine the equivalent number of points that the sensor is capable of addressing. Most commercially available systems are capable of addressing at least 300 points and the highest performing systems, up to 10000 points. Clearly, the sampling rate of the digitizing electronics must be suffident faithfully to replicate, in real time or by interleaving, the backscatter signals received. Further performance criteria include the resolution of the measurand, which is usually limited by signal-to-noise ratio at the receiver although some suppliers have chosen to define this performance parameter in relation to the resolution of their display. Finally, in systems where intensive digital averaging is normally relied on to improve the signal quality, the measurement time is an important measure of performance, since it determines the response time to a change in the measurand.

The spatial resolution depends on the duration of the probe pulse, the bandwidth of the fiber (at all relevant wavelengths), the bandwidth of the receiver and associated electronics as well as the sampling rate of the digitizing system. For constant probe power, as the spatial resolution, is increased the signal-to-noise ratio at the output of the receiver degrades. Thus, enhancing the spatial resolution usually results in a degraded resolution of the measurand (owing to a worse signal-to-noise ratio), a longer measurement time (owing to longer averaging times) or a shorter range to reduce the cumulative losses experienced by the signals. An example of distributed temperature sensing data is shown in Fig. 2.

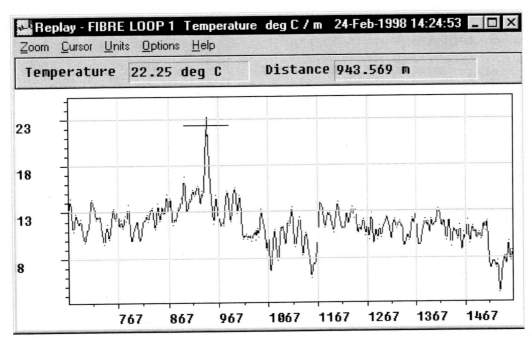

Figure 2: An example of data from a distributed temperature sensor

To date, most of the systems available commercially operate on multimode fiber, usually of the 50/125 graded-index type, although some of the equipment has been designed to operate on the larger 200/280 fiber type. In general, the preferred operating wavelength increases with the desired range. Thus short distance systems operating at 850 or 900nm based on semiconductor laser sources, are available with ranges of up to 1-3km; spatial resolutions of 1-3m are quoted. York Sensors' most widely installed product is based on a diode-pumped solid-state laser operating at 1064nm and achieves a spatial resolution of around 1m over up to 10km of fiber. For fibers lengths beyond 10km, the wavelength of the source is preferably increased to 1300nm or 1550nm. York Sensors has supplied some 250 DTS systems to date, most of these based on 1064nm sources.

As the operating wavelength is increased, a number of adverse factors affect the near-end performance of the system. Thus, increasing wavelengths result in reduced scattering factors that reduce the initial signal, but over long fiber distances may be compensated by reduced transmission losses. Moreover, the performance of the detectors degrades markedly beyond about 1000nm and therefore long-wavelength systems are unattractive for distances below about 10km.

2. DISTRIBUTED TEMPERATURE SENSING IN OIL & GAS APPLICATIONS

Distributed temperature sensing has found applications in a large number of fields, including electrical energy distribution[7], plant monitoring[8], geological measurements, fire detection, environmental monitoring and the construction industry. It has taken a little longer to become established as a proven technique in the oil & gas industry. The principle issues that required addressing before wide-scale adoption were; concerns over age-related attenuation and longevity of systems, practical sensor deployment methods and interpretation and demonstrating the value of distributed temperature data in oil gas production.

2.1. Concerns over attenuation and longevity of systems

Optical sensors do have some inherent drawbacks that have to be evaluated when they are considered as an alternative to conventional electronic transducers. In the case of optical distributed temperature sensing, it is crucial to avoid excessive build up of either intrinsic or extrinsic energy losses in the optical fiber. Intrinsic energy losses arise from the scattering and absorption of the light in the optical fiber medium. Connectors generate extrinsic energy losses as do field splices, and fibers that have tight bends and excessive heat or mechanical damage. Excessive losses will lead to a gradual degradation in measurement range or complete loss of signal in the extreme case.

The environment encountered in down-hole oil & gas production is relatively hostile and there are several factors that can give rise to either intrinsic or extrinsic energy losses.

2.1.1. Hydrogen (H$_2$) attenuation

The first of these is H$_2$ attenuation. It has been recognized since 1982[9] that there are absorptive losses associated with hydrogen that is dissolved in silica glasses. Due to its small size, hydrogen, which may be present around an optical fiber, can readily diffuse into the central light-guiding region of the fiber. It is within this silica-based core of an optical fiber that the major problems occur.

Two major types of optical loss effects have been attributed to hydrogen[10].

a) Absorption due H$_2$ to that diffuses into interstitial sites in the silica network. This phenomenon has been found to affect all silica-based glass fibers, both single and multimode.

The most significant effect is a strong absorption at 1.24 μm. Further peaks are also observed at longer wavelengths (a rising loss edge beyond about 1.5 μm). The diffusion of H$_2$ into the fiber core can be measured by the growth of the 1.24μm peak. As the concentration of H$_2$ increases the breadth of each absorption band also increases. Overall, these absorptions prevent the transmission of light at the required wavelength. The solubility of H$_2$ in bulk silica is linear with pressure; however, it decreases as temperature increases. The losses due to H$_2$ absorption are reversible. When external pressure and/or the H$_2$ source are removed the H$_2$ diffuses out of the core and the light transmission characteristics are restored with time.

b) Absorption due to chemical reaction of the diffused H$_2$ with the glass constituents to form, for example, hydroxyl (OH) groups. These have distinctive absorption bands that lead to high attenuation of the transmitted light.

Even for a pure silica fiber, the silica within the fiber core is not a pure, –Si–O–Si–O–, three-dimensional matrix. There remain some –Si=O sites, especially at the edge. Under certain conditions these can react with H$_2$ to produce 'dangling' – SI–OH sites. Such hydroxyl groups absorb strongly in the transmission region of interest (1.38, 0.95 and 0.72μm, with the strongest being at 1.38μm).

Not only does H$_2$ attack the silica core of the fiber, it can also chemically react with other species present, such as dopants. This can lead, for example, to the formation of Ge–OH which absorbs strongly at c. 1.4μm. Chemically induced attenuation is worse when both GeO$_2$ and P$_2$O$_5$ are present in the core of the fiber as the P$_2$O$_5$ catalyses the reaction to produce the Ge–OH species. Additionally, reaction of H$_2$ with GeO$_2$-doped fibers can also induce attenuation in the shorter wavelength region (<1μm).

H$_2$ can also react with defects within the core that contain alkali impurities. This is manifest in the growth of a long wavelength absorption edge. Additionally, H$_2$ can induce attenuation in undoped fibers (pure silica cores), for example those with a fluorine-doped cladding. In this case, absorptions due to HF (1.44μm) and SiH (1.53μm) can be detected. In single mode fibers, the evanescent field carried in the cladding will interact with these absorptions and result in increased attenuation of the fiber.

The intensity of all chemically induced absorptions grows with time (as the chemical reactions proceed) and they are irreversible. The rates of reaction are also accelerated at higher temperatures and increased H$_2$ concentration. Thus once it has been created, chemically induced attenuation cannot be removed.

Figure 3 shows a series of absorption spectra of an optical fiber with time. The experiment was carried out at Sensa and designed to study the effects optical attenuation in the fiber over time. The growth of the –OH and H$_2$ absorption peaks are clearly visible.

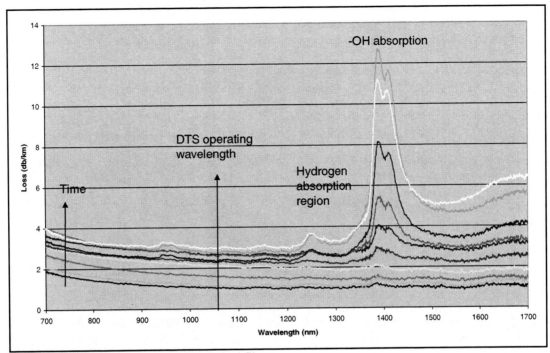

Fig. 3 Absorption of H_2 & OH into doped silica optical fibers

2.1.2. Potential sources of hydrogen

The ingress of H_2 into an optical fiber can induce considerable optical attenuation. However, H_2 will not diffuse through steel tubing at temperatures of up to 302°F (150°C) thus steel tubing can provide an impermeable barrier. Atomic hydrogen, H, will, however, diffuse readily through steel and thus can present a problem. H can be formed by the dissociation of H_2. This reaction is extremely slow up to 302°F (150°C) but may be significant at 572°F (300°C). H may therefore present a problem at elevated temperatures. H may recombine to H_2 after diffusion or attack the optical fiber itself. H_2 can either be present in the general environment around or within the optical fiber, or may occur when hydrogen sulphide (H_2S) is present.

In the former, H_2 exists in the atmosphere, but at less than 1ppm by volume. In an oil/gas environment the natural concentration may be significantly greater but will still be a relatively minor component. H_2 can also be formed by the electrolytic reaction between metallic elements in a cable, by water corrosion with metal components, and by the evolution of gasses from coating materials and their chemical reactions (decomposition).

H_2S is commonly found in oil/gas environments and can cause significant problems. Overall, H_2S levels of between 400 and 1000ppm by volume have been measured in the H_2S service environment. For example, H_2S in water is only mildly acidic but it will reduce iron causing deterioration. H_2S may also react with polymers present to weaken/degrade them, although normally not rapidly. H_2S is also the most common and potentially troublesome source of hydrogen. At atmospheric pressure H_2S reacts with 'steel' to produce atomic hydrogen, H. H_2S also hinders the combination reaction of H to H_2 and thus the atomic hydrogen is able to start to diffuse rapidly into the steel.

However, the problem of H_2 generation and diffusion is outweighed by another action of H_2S on steel at elevated temperature. At 302°F (150°C) in seawater, pitting will occur in steel. Only advanced alloys will resist. Pitting will cause perforation/total breakthrough through the steel tubing. If this occurs, then H_2 diffusion is not the major problem! The physical integrity of the optical fiber will be threatened. In contrast, if the correct corrosion resistant materials are selected to prevent corrosion and hydrogen-scavenging gel is used to collect environmental H_2, then there will be no H_2 diffusion into the fiber under conditions normally encountered in the oilfield.

2.1.3. Methods of preventing hydrogen ingress and hydrogen-induced attenuation

Three main methods have been developed to reduce the effect of H_2 on the optical transmission of optical fibers.

i). Altering the fiber dopant compositions to reduce chemically induced attenuation. Additives (getterers) can be included in the core. These preferentially react with H_2 to produce species that do not absorb in the transmission region.

ii). Redesigning the fiber cables to avoid the possibility of H_2 or H generation. The use of advanced alloys will prevent H being generated when H_2S is present at higher temperatures.

iii). Employing hermetic coatings on the fiber to block the diffusion of any H_2 that may be present (this will not mask the effect of H_2 present in the core at fiber manufacture which could cause small but significant long-term loss increases at elevated temperatures). This is the most efficient method, particularly at elevated temperature.

2.1.4. Hermetic coatings

Whereas the diffusion coefficient of H_2 is rapid in silica, it can be quite slow in materials such as metals, carbon and certain ceramics. For example, at 302°F (150°C) H_2 will not diffuse through steel (or if it does, then very slowly). Previous studies[9] on the predicted H_2 levels in hermetic and non-hermetic fibers have shown that the level of H_2 in the core of the hermetic fiber would be c. 10^{-4} atmospheres after 25 years at 70°F (21°C) with an external H_2 pressure of 1 atmosphere. A similar non-hermetic fiber would be saturated with H_2 in the core after as little as one week under similar conditions. The associated level of optical loss would thus also be significantly reduced for the hermetically jacketed fiber. Similarly it has been shown[9] that a carbon hermetic coating greatly reduced H_2 diffusion at temperatures up to 306°F (152°C) with an external H_2 pressure of up to 144 atmospheres.

The hermetic layer used on Sensa optical fibers is a carbon matrix prepared by chemical vapor deposition. Working in collaboration with our fiber manufacturer, it has been shown that increased resistance to H_2 diffusion has been achieved with an improved carbon matrix (E Lindholm, Lucent Specialty Fiber Technologies, Private communication). This greatly reduces the permeation of H_2 into the core. Thus this improved fiber, accompanied with steel encasement of the fiber, further reduces the risk of H_2 induced attenuation.

2.1.5. The effect of exposing fibers to liquids during deployment

The second factor to take into consideration when designing optical fiber systems for down-hole use is liquid ingress into the fiber. The ingress of liquids into optical fibers can lead to both increases in fiber attenuation and eventual mechanical failure of the fibers. In the case of the former, decomposition products passing into the fibers create absorptive losses in the waveguide and more significantly, cause mechanical damage to the fiber that leads to micro-bending effects. Water is extremely harmful to optical fibers. The presence of water, if permitted, will extend any surface cracks that may naturally occur in the fiber surface and these will be elongated by a stress cracking process. The mechanical strength of the optical fiber will be reduced to the point where it cannot support its own weight and failure will occur. Once again the use of hermetic coatings in situations where the optical fiber is likely to be exposed to water has been the traditional means of combating this problem.

Deployment of optical fibers by fiber-blowing techniques developed by the telecoms industry[11] requires that optical fibers are exposed to the presence of liquids at high pressure during this installation phase. Water is often the easiest fluid to use, but does increase the likelihood of optical losses occurring at elevated temperatures of greater than 302°F (150°C). Other fluids have been used to install fibers with blown-fiber apparatus for temperatures greater than this threshold with success. Over a period of 3 years a series of tests have been conducted in-house using special test apparatus that simulated down-hole conditions to guide incremental improvements to polyimide and hermetic carbon coatings. This work has resulted in both elevating the threshold at which water can be used for fiber-blowing and also reduce overall losses in fibers at temperatures >482 °F (250°C) when other special proprietary fluids have been used.

Figure 4 demonstrates the improvement that has been made in the elevated temperature performance of our fibers. The fibers were treated identically (subject to pressurized water prior to heating) and tested together[10]. The improved polyimide-coated fiber showed little or no increase in attenuation even at >572°F (300°C).

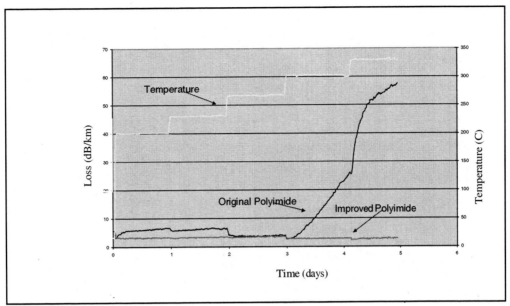

Fig.4 Advances in polyimide & hermetic carbon coatings resulting in improved high temperature performance (1999)

2.1.6. Micro-bending effects

Finally, optical fibers also suffer increased extrinsic energy losses due to micro-bending effects. Large bends of cable and fiber are macro-bends and small-scale bends are micro-bends. Macro-bends rarely create difficulties when optical fibers are deployed down-hole whereas care has to be taken to avoid forming micro-bends particularly at higher temperatures in the region of 482°F (250°C). Micro bending can occur when stresses are built up in the fiber coatings as either the fiber is heated or cooled. Probably cyclical heating and cooling offers the greatest challenge for optical fibers in respect of micro-bending losses. Tests performed for Canadian cyclic steam operations indicate that there is severe micro bending when the fiber is returned to ambient reservoir conditions after being exposed to temperatures as high as 644 °F (340°C). Further optical fiber coating development will be required to take measurements accurately across the full range of 86°F (30°C) - 644 °F (340°C). An illustration of the micro-bending problems exhibited by early optical coatings is show n in figure 5.

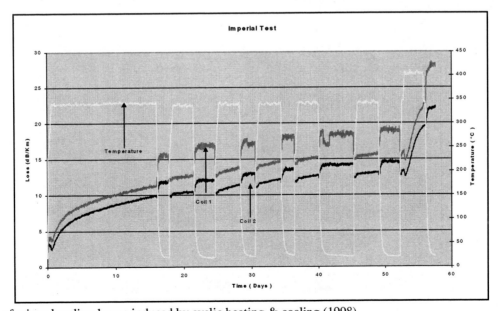

Fig. 5 Effects of micro-bending losses induced by cyclic heating & cooling (1998)

2.2.   Practical sensor deployment

One method of deploying optical sensors that has been developed utilises a Sensor Highway ™ conduit concept that allows thin filament sensors to be placed in the well utilising fluid drag. Coated optical fibers have been placed into pre-installed conduits in buildings for communication networks since 1991 using similar techniques[11]. To date over 40,000 km of thin-coated optical fiber having a diameter of a few hundred microns has been reliably installed for data-coms purposes in this manner. The conduit is placed either in a single length or in a u -tube form into the well at the time of completion and the optical fiber sensors can then be installed at any time using fluid drag. See Figure 6. Certain components in the well such as tubing hangers and packers have to be modified allow the conduit to straddle the hydrocarbon producing formation. Robust clamps are also required to secure the conduit to the production tubing.

This system offers the following advantages:

- cable dimensions are kept to a minimum.
- down-hole splices in the optical fiber are eliminated.
- sensors can be installed at any time after the well has been completed.
- sensors can be replaced if they are faulty.
- sensors can be upgraded with the life of the well.
- sensors can be "buffered" using specially formulated fluids that are placed in the conduit.

The system is becoming widely adopted in oil & gas production. Over 130 installations have been made in California, Canada, Venezuela, Indonesia and Europe. Well temperatures range from 160°F (70°C) to 550°F (287°C). The system is increasingly being used in horizontal wells where up to 10km of optical fiber are deployed by this technique in a loop. See Figure 7. an example of an extended reach horizontal well. Numerous horizontal wells have been logged with this technique. The fluid drag process easily transports the optical fiber to the toe of the well and back and eliminates the need to run coiled tubing to gather temperature data. When compared to wells where thermocouples are installed, the system also offers many more measurement points giving a comprehensive description of heat distribution in the well bore.

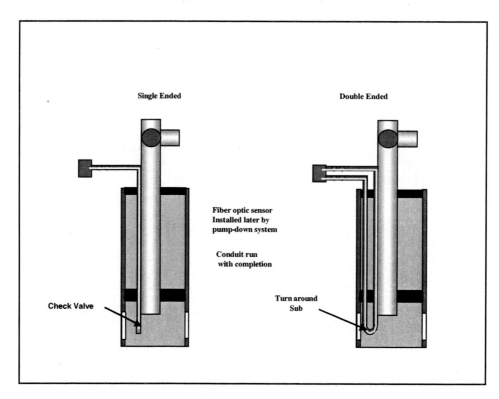

Fig. 6 Fiber installation using hydraulic drag

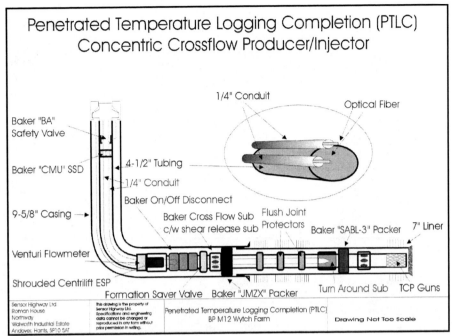

Fig.7 Extended Reach Horizontal well equipped with DTS

2.3.   Value of distributed temperature data in oil & gas wells

2.3.1.   Operating the distributed temperature sensor

A DTS system suitable for use in the oilfield comprises an opto-electronic unit containing the laser source and receiver and a portable PC. The sensor can be either a loop of optical fiber having both ends connected to the opto-electronic unit, or a single length of optical fiber with just one end connected. In either case the total length of the optical fiber can be up to 32,000 ft long.

During down-hole data acquisition the opto-electronic unit is coupled to the down-hole fiber via a junction box and cable near the wellhead. See Figure 8. Data acquisition software then manages the process of gathering Raman backscattered light and converting this to time-dated binary files of temperature data. This is most commonly then put in an Excel Spreadsheet for further analysis. There is a trade off between response time, spatial resolution, sample resolution and sensor length. Generally speaking, the longer the measurement time, the better the resolution of the data. For ex ample, a well in shallow Californian steamflood fields with approximately 2,000 ft of single ended optical fiber, a measurement time of 4 minutes 30 seconds results in an accuracy of 0.5 °F and resolution of 0.2 °F with a spatial resolution of 1 meter. Alternatively in a long horizontal offshore well equipped with a double-ended loop of fiber, the measurement time is typically 60 mins to give the same results. This gives a data point every meter point along the well bore from wellhead to the bottom of the well.

More recently the DTS Systems are being permanently installed at the well-site to allow continuous acquisition and recording of down-hole data. Recent advances of internet technology now allow the added utility of automatic data transfer from remote locations to server-based databases allowing easy data access and downloads.

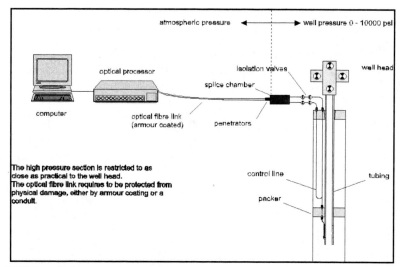

Fig. 8 Operation of the DTS for oil & gas well monitoring

2.3.2.   Permanent down-hole monitoring with the DTS

Permanent down-hole monitoring can provide valuable information for production decisions without the need to perform intervention into the well to gather data. The acquisition of distributed temperature data over the reservoir reveal trends,which, when analyzed, help corroborate reservoir inflow and well performance characteristics. Trend analysis can be used to help identify and in some cases manage the following down-hole conditions[12].

- flow contribution across zones and long horizontal sections
- deduction of mass flow rate
- identification of water and gas break through zones
- optimization of flow to mitigate water and gas breakthrough
- damage or noncontributing intervals
- variances in permeability

In addition to monitoring across the reservoir, having a continuous sensor along the entire length of the well enables the collection of valuable data to manage:

- gaslift valve operation
- electrical submersible pump (ESP) performance
- identification of hydrate formation

The DTS systems have most value when used in the newer generation of horizontal and multi-lateral wells where logging costs using conventional wireline tools can be prohibitively high.

2.3.3.   Using a thermal well bore simulator to model and interpret DTS data

In 1962 Ramey, one of the early pioneers of advanced well logging techniques published a method of determining flow in well bores from temperature measurements[13]. His solution approximates a steady-state heat transfer in the well bore and unsteady radial heat conduction to the surrounding earth for single-phase flow. However this is applicable to vertical and deviated wells only. Li & Zhu[14] extended the concept by proposing a two-phase model for analyzing flow from production log temperature and flow meter data based on mass and energy balance equations in vertical pipes. However many wells are now completed horizontally and in horizontal pipes the effect of friction can have a significant effect on the observed temperature profile and must be taken into account[15.] In 1996 Brown[16] recognized the problems of using conventional production log analysis in undulating horizontal wells. The spinner flow meter response is significantly distorted by the effect of two-phase stratification and also the cost of the operation is considerably increased by the requirement to run logs on coiled tubing. Consequently a method of permanent monitoring that can identify inflow, the effects of different fluid phases and responds to production both inside and outside the casing is a realistic alternative to spinner logs in horizontal wells.

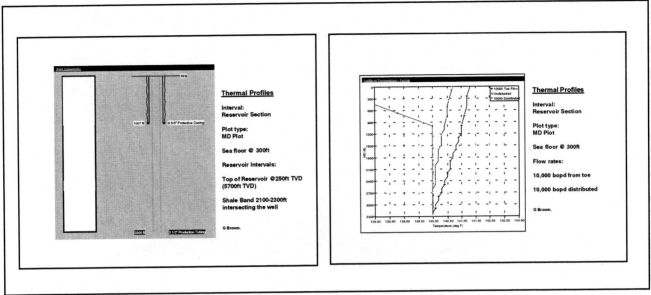

Fig. 9 Example of a thermal well bore simulation model

Software is now available[17] to analyze distributed temperature data and solves the energy equation for well bore and formation and the equations of mass, momentum and energy for each flow stream in a multi-stream model of steady state or transient flow in horizontal wells. This allows temperature profiles to be computed based on estimates of three-phase flow for a multi zone undulating horizontal well. Whilst a single unique solution may not be always achieved using temperature alone, a thermal well bore simulator allows a variety of producing scenarios to be examined so that the most likely solution, which fits both the observed temperature response and other reservoir and production data, can be determined.

2.3.4. An example of applying the well bore simulator technique to an extended reach horizontal well

A number of extended reach horizontal wells have been equipped with the system at BP's Wytch Farm oil field located on the south coast of Great Britain. The second of these drilled in 1999 exhibited temperature profiles that were distinctly abnormal.

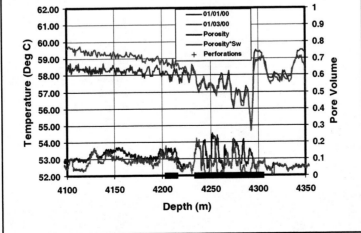

Fig. 10 M-17 Distributed temperature data showing seawater break-through and formation water break-through

Significant cooling at the end of the well was attributed to seawater break through from a nearby seawater injector. See Figure 10. If we assume that the break though of water in the lower interval is initially the only water production, its rate is known from surface measurements to approximately 500 bpd. Using this as a starting point the thermal well bore simulator can be used to match the recorded temperature profiles at other points in the reservoir by adjusting flow rate and water/oil content. See Figure 11. This approach assumed that all the other inflow took place at the geothermal temperature. The analysis was performed on both 1st January and 1st March 2000 temperature data corresponding to 25% and 35% water production.

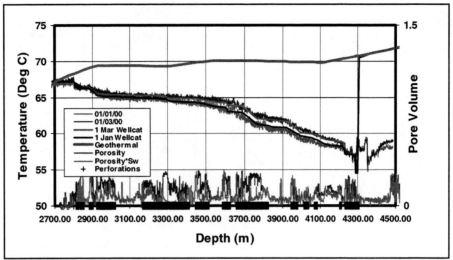

Fig.11 Matching well response using the thermal well bore simulator

The increase in temperature at the top of the lowest interval is primarily achieved by switching from oil to water production in the model, which changes the thermal properties of the produced fluid, suggesting that this interval is the cause of the increase in water production over the time period. By matching the thermal simulator model to the recorded temperature data, producing zones can be identified and estimates of production obtained. See Figure 12.

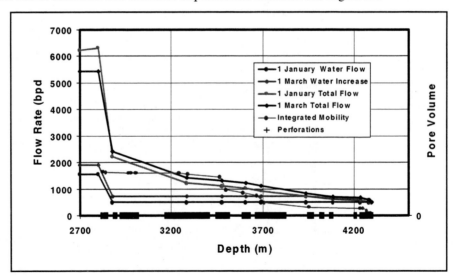

Fig. 12 Flowing profiles in the M-17 extended reach horizontal well

2.3.5. System reliability

Since 1995 over 130 wells have been equipped with the distributed temperature sensor in North and South America, Europe and the Far East. System performance has been excellent with 90% of systems operating. There have been no failures below 300°F (149°C) since the start of operations in 1996 and no failures at all during the last two years. Two types of failure modes have been experienced at high temperatures in steamflood operations. The first was due to liquid ingress into the fibers in Canada in 1996. The improvements made to fiber-blowing fluid selection and fiber coatings have solved this. Failures occurring in California in 1997 and 1998 have been due to subsidence in the oil producing reservoirs leading to collapse or buckling of the production tubing. This has led to total failure of the well including fiber optic cables[18]. The latter type of failure, the statistics for which are included in the non-operating category in the data of Fig. 13, are not fundamental to the fluid-drag deployment process and are a consequence only of the local geological conditions in which the fibers were installed

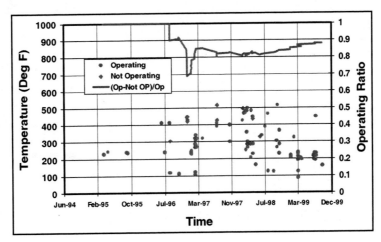

Fig. 13 System reliability 1995 – 1999

2.3.6. System expansion

In 2000 and 2001, extrinsic Fabry-Perot interferometric (EFPI) pressure sensors and acoustic sensors based on fiber optic time-domain-reflectometry[19,20] will be added to the DTS system in the well. Although these additional sensors will operate on a quasi-distributed or multi-point basis they will be useful in assisting in the interpretation of the temperature profile in the well. Pressure sensors will assist in determining reservoir heterogeneity and permeability, allow transient pressure analysis and assist in identifying fluid types in the wellbore. In "Smart" or "Intelligent" wells they will allow effective balancing and commingling of different layers or reservoirs. Acoustic sensors will determine single-phase flow rate, indicate changes and type of flow regime as well as identifying the origin of sand production. All of these sensors shall gather data using a single multi-fiber cable that is installed into the well.

It is envisaged that the complete suite of fiber optic sensors shall operate reliably for the life of the well and eliminate the need to perform costly intervention to gather data.

3. DISTRIBUTED SENSING WITH SPOTANEOUS BRILLOUIN SCATTERING

The monitoring of deep-water, long step-out, wells will challenge the current performance of Raman based systems[19]. Raman DTS technology, which is capable of a performance in the order of 1°C rms at 30km with 10m resolution in 600s, is approaching the ultimate physical limits imposed by receiver sensitivity and maximum launch power before the onset of non-linear effects. Our studies have shown that in order to be a an effective long-range well logging device the long range DTS system will require significantly improved performance than Raman systems can currently offer. That is to say the systems must offer resolution in the range of 0.1°C rms over distances exceeding 30km and spatial resolutions of order 3-5m.

Brillouin scattering, especially the Landau-Placzek ratio measurement[21,22], is being considered as an alternative. Although the current performance of Brillouin systems is still comparable with that of Raman systems[19], there is still considerable scope for further improvements and this technology may therefore offer the monitoring solution required to unlock the potential of deepwater reservoirs.

The principle of the BOTDR DTS is based on the Landau Plazcek ratio where the temperature insensitive Rayleigh signal provides a reference measurement of the fiber background losses and the temperature sensitive Brillouin signal provides a measurement of the temperature. In the results shown in Fig. 14, a narrow-band fiber Bragg grating filter (FBGF) was used[23] that achieves recovery of the Brillouin signal by suppressing the Rayleigh signal. In this way, the Brillouin light path is subject to minimum attenuation and is frequency independent.

These results demonstrate a temperature resolution of <7°C over a 25km length of fiber with 2m spatial resolution in a 600s data acquisition time. It is also shown that if the measurement time is extended to 3hou rs then a temperature resolution of <1°C is possible over most of this distance rising to ~1.7°C at 25km. These results are illustrated in Figure 14.

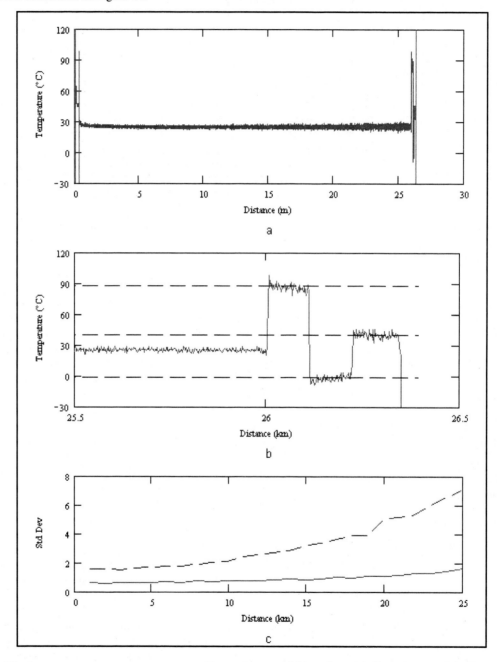

Figure 14. Brillouin temperature measurement over 3hour data acquisition time. (a) Complete sensing fib er, (b) temperature calibration unit at far end, (c) Standard deviation of temperature measurement along fiber compared with that for the 10min data acquisition time (dashed).

4. CONCLUSIONS

Distributed temperature sensing has been found to be an effective way of monitoring the production characteristics of oil wells once cost effective and reliable techniques were developed to install optical fibers in the hostile conditions present in oil & gas production. Improvements have been made to hermetic coatings to withstand the effects of being subjected to high temperatures and pressures when utilized down-hole. In order to develop improved coating technology and a robust cable design, the effects of hydrogen (H_2) and atomic hydrogen (H) have been well understood in the laboratory but there has been no evidence of hydrogen darkening across the 130 systems installed by Sensa since 1996.

DTS today is the only system that can offer a total wellhead to well-total-depth data profile that identifies flow patterns, changes in fluid properties and monitors the overall integrity of the wellbore and artificial lift equipment without intervention into the well. The data sampling frequency given by the 1meter spatial resolution is such that the sensor does not have to be accurately positioned in the well.

DTS systems have been installed in a wide variety of types of oil well ranging from shallow high temperature steamflood wells to extended reach horizontal wells. The system is also now being widely perceived as "SMART" technology that will form the backbone of a new type of "Intelligent Wells". It has also been demonstrated that tools are readily available to model and analyze distributed temperature data and conventional wireline logging intervention can be reduced or eliminated.

The addition of quasi-distributed pressure and acoustic sensors will add further utility to the system and broaden its ability to fully describe well and reservoir behaviour. Future development of distributed sensing in the upstream oil industry include extending the range of operation of the systems to allow measurements to be made in remote subsea satellite wells that can be greater than 30km from a host platform of floating production facility. It is most likely that these systems will be based on Brillouin scattering that offers both distributed temperature and strain measurements. The ability to measure distributed strain will have great value in protecting the integrity of subsea flowlines and risers. Distributed optical sensing has the potential to gather data of a magnitude and accuracy and operate at levels of high reliability not previously thought possible. It will allow the management and optimization of energy from along the entire producing system comprising bottom of the well to platform, refinery and beyond.

ACKNOWLEDGEMENTS

The authors would like to thank BP and it's Wytch Farm c-venturer's; Premier Oil plc, ONEPM Ltd, Kerr McGee North Sea (U.K.) Ltd., and Talisman North Sea Ltd. for permission to publish the data in this paper. We also acknowledge Peter Wait of York Sensors and the support provided by a LINK scheme in collaboration with University of Southampton (US) and Pirelli Cables Ltd (PC). One of us (AHH) is also grateful for the useful discussions with T. P. Newson, S. M. Maughan, H. H. Kee (US) and R Pragnell (PC).

REFERENCES

1. Hartog, A.H., 1983, "Distributed Temperature Sensor Based on Liquid-Core Optical Fibers" J. Lightwave Technol., LT-1:498-509.
2. Barnoski, M.K. and Jensen, S.M., 1976,"Fiber-Waveguides: a Novel Technique for Investigating Attenuation Characteristics", Appl. Opt., 15:2112-2115.
3. Dakin, J.P. et al., 1985, "Distributed anti-Stokes Raman Thermometry" Proc. 3rd Int. Conf. on Opt. Fiber Sensors, San Diego, February (post-deadline paper)
4. Hartog, A.H. et al., 1985, "Distributed Temperature Sensing in Solid Core Fibers" Electron. Lett., 21:1061-3.
5. Farries, M.C and Rogers, A.J, 1984, Distributed Sensing Using Stimulated Raman Interaction in an Optical Fiber, Proc. 2nd Int. Conf. on Optical Fiber Sensors, Stuttgart, paper 4.5, pp121-32.
6. Dakin, J.P., 1987, " Distributed Fiber Temperature Sensor using the Optical Kerr Effect", Proc. SPIE, 798, Fiber Optics Sensors II, pp149-156.
7. Hartog, A.H., 1995, "Distributed fiber-optic temperature sensors: technology and applications in the power industry", Power Engineering, June issue.

8. Hartog, A. H. 1995, "Fiber-optic temperature sensors monitor Wakamatsu" Modern Power Systems, February issue, pp25-28.
9. Lemaire. P. "Reliability of optical fibers exposed to hydrogen: prediction of long-term loss increases."
10. Hawthorne W.H, The effect of hydrogen on the optical transmission of optical fiber. Internal Report Sensa 2000.
11. BT Telecommunications plc, US Patent 5022634 & 5199689.
12. G.Mckay, P.W. Bixenman, G.Watson, Advanced Sand Control Completion with Permanent Monitoring, SPE 62954 2000 Annual SPE Technical Conference and Exhibition, Dallas.
13. H.J. Ramey. "Well bore heat conduction" SPE96 J. Pet Tech. (April 1962).
14. H. Li, D. Zhu, L.W. Lake annd A.D. Hill. "A new method to interpret two-phase profiles from the temperature and flowmeter logs" SPE 56793. 1999 Annual SPE Technical Conference and Exhibition, Houston.
15. Brown, G. Kennedy, B. Meling, T. "Using fiber-optic distributed temperature measurements to provide real-time reservoir surveillance data on the Wytch Farm Field horizontal extended reach wells. SPE 62952, 2000 Annual SPE Technical Conference and Exhibition, Dallas.
16. Brown, G., Toro, M, Rees, H. "Using production logs in horizontal wells" 1996 SPWLA Annual Technical Conference, Paris.
17. Landmark Graphics Wellcat-Prod^{TM} Extended Reach Well Model, Beta Version software.
18. Carnahan, B. Clanton, R. Koehler, K. Williams, G. "Fiber Optic Monitoring Technology". SPE 54599, 1999 Western Regional Meeting, Anchorage, Alaska.
19. Hartog, A.H. in "Optical Fiber Sensor Technology", Grattan, K.T.V. and Meggitt, B.V. (Eds.), Chapman and Hall, London 1995, ISBN 0 412 59219 X
20. J.P. Dakin, C.A. Wade, M.L. Henning: Electron. Lett. vol. 20, pp 53-54 (1984).
21. P. C. Wait and T. P. Newson, "Landau Placzek ratio applied to distributed fiber sensing", Optics Communications, vol. 122, pp. 141-146, 1996.
22. T. R. Parker, M. Farhadiroushan, V. A. Handerek and A. J. Rogers, "A fully distributed simultaneous strain and temperature sensor using spontaneous Brillouin backscatter", IEEE Photonics Technology Letters, vol.9, pp.979-981, 1997.
23. Wait, P.C. and Hartog, A.H. "Spontaneous Brillouin based distributed temperature sensor utilising a fibre Bragg grating filter for the separation of the Brillouin signal". Submitted for publication. "

Large scale multiplexed fibre-optic arrays for geophysical applications

Philip J. Nash, Geoffrey A. Cranch and David J. Hill

DERA-Winfrith, Winfrith Technology Centre, Winfrith Newburgh, Dorchester, Dorset, DT2 8XJ,U.K.

ABSTRACT

Fibre optic sensors are becoming a well-established technology for a range of geophysical applications, and static pressure and temperature sensors in particular are now comparatively well developed. However, rather less attention has been paid to systems for measuring dynamic quantities such as acoustic and seismic signals. Furthermore, the very large multiplexing potential of fibre optic sensing systems has yet to be fully explored for geophysical applications. However, development of fibre optic sonar systems for military applications has proven the viability of large multiplexed arrays, and demonstrated advantages which include electrically passive arrays, long term reliability and the potential for operation in very deep (>3000m) water. This paper describes the applications for large scale fibre optic sensing arrays in geophysical metrology. The main applications considered here are ocean bottom cables and streamers for marine seismic, and downwell seismic systems. Systems can require up to several thousand channels and the use of multi-component sensors, which include 3–axis geophones and hydrophones. The paper discusses the specific requirements for each application, and shows how these requirements can be met using a system approach based on time and wavelength multiplexing of interferometric sensors. Experimental and theoretical studies at DERA into the performance of highly multiplexed systems are also described, together with initial development work on fibre optic hydrophones and geophones.

Keywords: optical hydrophone, optical geophone, multiplexing, DWDM, TDM, interferometric, seismic sensing

1. INTRODUCTION

One of the major breakthroughs in the application of fibre optic sensors in recent years has been in the oil/gas industry. It has been recognised by the industry that fibre optic sensors have a major role to play in the rapidly growing area of reservoir characterisation[1]. It has been shown that production yields from oil/gas reservoirs can be greatly improved by having a better understanding of the oilfield environment. This can be obtained through use of a range of sensors, which can be either inserted into wells or deployed in the vicinity. Sensors deployed into wells must survive the harsh environment, which includes temperatures of up to 200 ^{0}C and pressures of 20 000 psi. Existing electronic sensors are prone to failure in this environment, but fibre optic sensors offer the possibility of greatly increased reliability and an improved multiplexing capability. Fibre optic sensors for the measurement of temperature, pressure and flow have been successfully demonstrated, and are already beginning to be installed in production wells[1]. Technologies employed for these sensors have included in-fibre Bragg gratings[2], Fabry-Perot sensors[3] and distributed Raman and Brillouin sensors[2].

All the sensors described above are designed to measure static or quasi-static quantities i.e. the measurands vary slowly with time. However, interest is now turning to the development of fibre optic sensors for measurement of dynamic quantities such as sound and vibration. In this context, we use dynamic sensors to mean those sensors which measure time varying signals with a bandwidth of greater than 10 Hz. Of particular interest are hydrophones and geophones for the measurement of seismic and acoustic signals, both downwell, in the marine environment and on land. These applications require a quite different approach to both sensor and interrogation system design. For instance, one of the principal challenges with static sensors is the reduction of drift in order to achieve high long-term accuracy. For the dynamic sensors described here, the main requirement is to measure the amplitude and frequency of the signal variation rather than its absolute value. For instance, a hydrophone will not in general give any measurement of the absolute pressure. Emphasis is therefore placed on the achievement of wide frequency coverage with high dynamic range and low signal distortion. In addition, for most practical applications, systems will require large numbers of sensors and therefore an efficient multiplexing architecture. Many of these issues have already been addressed in the development of fibre optic sonar arrays for military applications. This paper discusses the specific requirements for highly multiplexed sensor arrays for geophysical applications, and the challenges that these requirements present. A system solution based on DERA's fibre optic hydrophone technology, developed for naval applications, is described, and the development required to make this

Industrial Sensing Systems, Anbo Wang, Eric Udd, Editors,
Proceedings of SPIE Vol. 4202 (2000) © 2000 SPIE · 0277-786X/00/$15.00

technology suitable for oilwell applications is discussed. Both system and sensors design issues are addressed, and the status of hydrophone and geophone technology is reviewed.

2. FIBRE OPTIC SENSORS FOR MEASUREMENT OF SOUND AND VIBRATION

The major types of dynamic sensor required by the oil industry are likely to be hydrophones and geophones. In this context, a hydrophone can defined as a device for the measurement of the dynamic pressure associated with an acoustic signal, and a geophone as a device for the measurement of vibration (in practice, this can either be an accelerometer or a displacement sensor). Both hydrophones and geophones have a wide range of applications in the oil industry, which can be divided into three main classes depending on the region of deployment; downwell, surface and marine. The applications are discussed briefly below and details of the requirements for each application are given in section 3.

2.1 Downwell applications
In these applications, sensors are introduced directly into the oilwell, and can be attached to the central tubing, the outer casing, or suspended in between, as shown in figure 1. Downwell applications for hydrophones/geophones can be divided into those which involve direct measurement of the local acoustic field, and those which involve measurement of external seismic signals. Measurement of the local acoustic field may yield information about flow rates within the central tubing and the performance of local machinery e.g. pumps. This can be achieved either by using geophones (vibration sensors) attached directly to the central tube, or by hydrophones suspended between the tube and the casing.

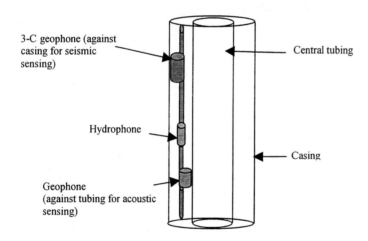

Figure 1: Downwell sensors for measurement of sound and vibration

Measurement of seismic signals originating outside the well can be used to build up a picture of the overall reservoir structure. This can be achieved either using vertical seismic profiling, crosswell seismic or by using microseismic signals generated continuously as the reservoir structure changes with time. The seismic sensors either comprise 3-axis geophones attached to the outside casing, hydrophones between the casing and the tubing, or a combination of the two (known as 4-component or 4-C sensing). At present, seismic measurements are normally carried out by inserting temporary wireline logging systems into wells, but the current drive is towards permanently installed systems which can be regularly monitored and will allow changing reservoir characteristics to be determined throughout the lifetime of the well (this is known as 4-D seismic, corresponding to three special dimensions and one time dimension). For the seismic application, a vertical array of sensors, probably between 10 and 100, is required. It will also be apparent that it may be difficult to distinguish between local acoustic and remote seismic signals. This may be achieved by positioning and/or of the sensor so that they are shielded form the unwanted signal. The use of 4-C devices will also allow discrimination between pressure and vibration. However, the dynamics of acoustic and seismic interactions within oilwells is not well understood and requires much further investigation..

For a fibre optic system, the objective is to locate all the electronics in a readily accessible surface unit, with an electrically passive all-optical downwell array. For both acoustic and seismic applications, the main perceived advantages driving the use of fibre optic sensors are:

- Removal of all electronics from downwell arrays leading to increased reliability at high temperature
- Efficient sensor multiplexing schemes allowing large numbers of sensors to be passively multiplexed
- Reduced array diameter

2.2 Surface applications
The major surface application is geophone arrays for seismic surveys. In this case, a large grid of sensors is normally used for exploration and initial oilfield characterisation. Typically 3-component geophones are used, spaced over an area which may be tens of kilometres in extent. Eventually, such systems may also be permanently installed for 4-D seismic applications.

The main advantages of fibre optic systems in this application are:
- Removal of electronics from the array, leading to increased sensor reliability and reduction in cable size
- Removal of requirement for batteries at sensor packages, leading to simpler deployment methods

2.3 Marine applications
At present, the main applications are in exploration seismic surveying. Two types of system are used: towed streamers and seabed laid ocean bottom cables (OBCs). Streamers comprise a number of parallel lines of hydrophones towed behind a specialist survey vessel, and can utilise several thousand hydrophone channels, with a streamer length of over 8 km. The latest generation of OBCs use 4-C sensor packages and can be up to 12 km in length. They are typically laid on the seabed in shallow water (<500m) in a grid configuration.

Although marine seismic systems have traditionally been used for exploration purposes, as in the other applications there is a strong drive towards permanently installed seabed arrays for 4-D reservoir monitoring. This marks a change in emphasis with seismic systems away from exploration and towards optimisation of the production process. Such permanent systems may be trenched, and form one element of a monitoring network which will also include a number of downwell arrays.

The main advantages of fibre optic systems for marine applications are as follows:
- Increased reliability of underwater arrays due to removal of electronics and reduced susceptibility to leak-induced failure
- Reduced size and weight of cables which will allow deployment of OBCs and streamers from smaller vessels
- Increased operating depth for OBCs due to reduced size and weight of cables

3.   APPLICATION REQUIREMENTS

It will be clear from section 2 that there are a wide range of applications for dynamic fibre optic sensor arrays in the oil industry, and the requirements for the different applications are likely to differ significantly. Even for a single application, the requirements may differ according to the precise circumstances and the geographical location. Furthermore, for some of the newer applications the requirements are still unclear. Table 1 summarises the typical performance requirements for the applications described above. The values quoted are meant to be simply an indication of the possible range of requirements for each application. In many cases, more precise values are either undefined or will be considered commercially sensitive.

Application	Sensor Type	No. sensors	Array size	Frequency range	Noise floor	Dynamic range	Distortion
Downwell seismic	4C	10-100	10m-1000m	1-500 Hz	10 nG/√Hz (geophone) at 10 Hz	>120 dB	< -50 dB
Downwell acoustic	1C geophone or hydrophone	1-10	1-100m	5-5000 Hz	1 mPa/√Hz at 100 Hz	>80 dB	< -30 dB
Land seismic	3C	1000-10000	10 km by 10 km	1-500 Hz	10 nG/√Hz at 10 Hz	>120 dB	< -50 dB
Streamer	Hydrophone	1000-10000	8 x 12 km arrays	1-500 Hz	1 mPa/√Hz at 10 Hz	>120 dB	< -50 dB
OBC	4C	100-1000	10 km	1-500 Hz	10 nG/√Hz (geophone) at 10 Hz	>120 dB	< -50 dB

Table 1: Typical seismic requirements for geophysical applications

Some of the parameters listed above require further definition:

Number of sensors: The number quoted is for the number of sensor packages. For a 4C sensor package, this may comprise 4 separate sensors within a common package, and therefore 4 separate optical channels.
System noise floor: Equivalent to minimum detectable signal. Can be specified either as a power spectral density (/Hz) or as an integrated noise power across the frequency band of interest. The PSD will normally be expressed in Pa/Hz for pressure sensors or nG/Hz for accelerometers.
Dynamic range: Here defined as the ratio (in dBs) between the system noise floor and the maximum signal level for which the specified distortion performance is achieved
Distortion: Normally expressed as total harmonic distortion

It has already been noted that the requirements for the various applications vary widely. However, a few general observations can be made.

Array size/number of channels - The size of array required varies significantly, with the downwell arrays being the smallest and the streamers probably the largest. However, with the exception of some of the acoustic applications, even the downwell systems are likely to require a minimum of ten 4-C packages (which will generally require 40 optical channels). In order to minimise requirements for cables and connectors, it is desirable to multiplex as many channels as possible onto a single fibre. Furthermore, to fully realise the advantages of optical systems, this multiplexing should be achieved without any electronics or electrical power within the array.

Frequency range - The frequency range of interest is generally lower for seismic than for acoustic applications: Existing systems will typically operate down to 10 Hz, but there is interest in extending this down to 1 Hz. In seismic applications, most of the energy is concentrated below 500 Hz, although some information is present at higher frequencies.

System noise floor - The system noise floor requirement is generally set to be below the background noise level in the environment of interest. For downwell applications, this will be set by a combination of background seismic and in-well acoustics, while for streamers it is normally set by flow noise. For OBCs, it will normally be set by background ocean noise levels. Although these levels vary considerably, in deep water and for the calmest sea conditions, the Deep Sea State Zero standard noise curve can be used. One of the most demanding requirements for seismic sensors is achievement of the specified noise level at very low frequencies.

Dynamic range/distortion - The dynamic range requirement is especially demanding for seismic sensors, because they need to be able to handle both the very large direct output from a seismic source, and the much smaller return signals. Furthermore, this needs to be achieved with very low distortion and inter-channel crosstalk. To achieve these demanding performance requirements in a highly multiplexed system is one of the key goals in proving the feasibility of fibre optic technology for these applications. Achievement of these goals requires an integrated systems approach which considers both the sensor, the multiplexing scheme and the electronic interrogation system.

Another factor is the environmental conditions in which the arrays must operate. Table 2 summarises the likely environmental conditions encountered. While each environment has its particular dangers, the downwell environment is probably the most hazardous. The combination of high temperature and pressure with a range of potentially damaging chemicals requires very careful attention being paid to packaging of sensors, connecting fibres and other components[1].

Environment	Temperature	Pressure	Chemicals	Other dangers
Downwell	Typically 150 ^{0}C (max 200 ^{0}C)	10000 – 20000 psi	H_2S, H_2, HCL, HF, CO_2	Mechanical deployment into well
Land	-40 to +80 ^{0}C	Atmospheric	None	Ultraviolet, handling hazards, traffic/animals
Marine	0 to + 40 ^{0}C	1500 - 7500 psi	Seawater	Mechanical stresses during deployment/recovery, Shark bite

Table 2: Typical environmental requirements for geophysical applications

4. FIBRE OPTIC HYDROPHONE SYSTEMS

Fibre optic hydrophones were one of the very earliest types of fibre optic sensor to be investigated, and have now been under development for military applications for over 20 years[4]. The technology has now developed to the stage where fibre optic based systems are serious candidates for future military sonar systems. Applications include towed arrays, hull-mounted arrays for submarines and surface ships, and deployable seabed arrays. In recent years, emphasis has been placed on the development of large, highly multiplexed arrays which achieve high acoustic performance and are robust enough to survive the marine environment.

Fibre optic hydrophones have been in development in a number of countries, including the Netherlands, Norway, France, Italy, and Japan. However, the major effort has probably been concentrated at the Naval Research Laboratory (NRL) and Litton Industries in the US[5] and at the Defence Evaluation and Research Agency (DERA) in the UK. Both NRL and DERA have concentrated on interferometric based sensing approaches, which give the highest sensitivity and can be readily multiplexed. NRL, working with Litton, have been involved in the development of systems based on frequency division and time division multiplexing (TDM). More recently, systems based on a combination of time and wavelength division multiplexing have been under development. In parallel with this, DERA have been developing a TDM/WDM approach based on reflectometric interferometry, which is described in the next section.

Many of the requirements for military fibre optic sonar systems are similar to those for oilfield applications. The requirement is for large arrays (100-1000+ hydrophone channels) with a frequency bandwidth of several kilohertz and a dynamic range of around 120 dB. Fibre optic hydrophone systems are therefore well suited to adaptation for oilfield applications. However, there are a number of key differences between military and oilfield (in particular, seismic) applications. Seismic systems are generally optimised for the measurement of pulsed signals over a large amplitude range while military systems are more typically (but not exclusively) concerned with the measurement of small continuous acoustic signals. Also, array sizes tend to be larger and frequencies of interest lower for seismic applications. Finally, military systems are almost exclusively based on hydrophones only. In order to transition fibre optic sensor technology into seismic applications, these differences must be addressed.

5. TIME AND WAVELENGTH DIVISION MULTIPLEXING ARCHITECTURES

The DERA reflectometric TDM architecture was originally developed in the 1980s by Plessey[6], and has been implemented in hydrophone systems with up to 32 channels. In combination with more recently developed WDM schemes, this forms the basis for multiplexing architectures which can efficiently combine over 200 channels onto a single fibre. Although developed originally for military applications, these architectures are highly suited for use in the applications discussed in sections 2 and 3.

5.1 TDM architecture principles

Interferometric fibre optic hydrophones and geophones operate by converting an acoustic or seismic signal into a strain in a coil of optical fibre. This imposes a phase change in an optical signal propagating through the coil, due to a combination of the physical length change in the fibre and the stress-optic effect (causing a change in refractive index of the fibre). This phase change is detected by beating this signal with a reference signal of a slightly different frequency which, when mixed on the photodiode, produces a beat frequency, or heterodyne carrier, equal to the difference in frequency of these two signals. The acoustic signal will appear as a phase modulation on this carrier and, in practice, the accuracy to which this modulation can be extracted usually determines the overall performance of the sensor. By pulsing the input signal, time of flight can be used to distinguish between different sensors. The basic TDM architecture is shown in figure 2. The hydrophone sensors, consist of a coil of fibre wound onto an air-backed mandrel. These coils are spliced serially together, separated by a directional coupler with a reflective mirror attached to one port. The other port of the coupler is index matched such that reflection only occurs in one direction, ensuring multi-path reflections are suppressed. The array module is interrogated with two optical pulses which are frequency shifted relative to each other and separated in time by a period set to twice the transit time in the sensor fibre. The return optical signal from the sensor contains an overlap of the two pulses where one has reflected from the first mirror forming the reference and the other has travelled through the sensor and back. On the photodiode, the overlapping pulses generate the heterodyne carrier signal that is phase modulated when a strain is applied to the sensor fibre. The heterodyne pulse from a single sensor is separated from those from the other sensors with a gate switch and this is then band pass filtered (BPF) and phase demodulated.

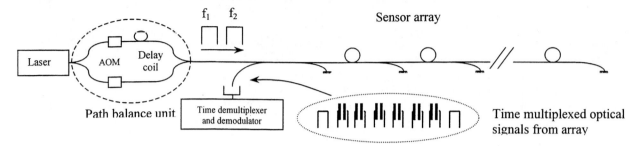

Figure. 2 : TDM architecture of optical hydrophones in an array module.

The optical power received from each sensor is equalised by tailoring the coupler power split ratios such that the coupler ratio of the Nth coupler, working from right to left in fig. 1a, is 1/N assuming no splice or coupler excess losses. The optical pulses are generated from Acousto-Optic modulators (AOM) which operate as both a switch and a frequency shifter. This architecture has the advantage of reducing the effective path imbalance in each of the sensor interferometers to below 1m which reduces the impact of laser frequency noise on the overall system noise, since the two pulses are generated from the same segment of light. In a practical system, eight sensors are typically multiplexed onto a single fibre and these sections can be arranged in a tree-type structure so that thirty-two are addressable through a single fibre pair.

5.2 TDM/WDM architecture

The level of multiplexing that can be achieved using TDM alone is limited by a combination of power budget and sensor sampling frequency considerations. For the major seismic applications discussed here, a higher level of multiplexing is desirable. Wavelength division multiplexing is a technique which has previously been used to increase multiplexing gain in TDM systems, but in the last few years advances in WDM component technology for telecommunications applications has led to the possibility of much more efficient multiplexing architectures. DERA have been developing a system architecture based on the use of dense WDM components, which has been previously reported[7]. An example of this architecture is shown in figure 2. In this arrangement, each TDM set of hydrophones, configured as described in sub-section 5.1, forms a module within a larger array, and each module is interrogated using a single optical wavelength. The multiple wavelengths required to interrogate the full array are launched into a telemetry fibre, and individual wavelengths are coupled out of this fibre into the TDM modules using optical add/drop multiplexers, which also couple the returning reflections from the module back onto the telemetry fibre. At the array output, the different wavelengths are routed onto separate photodiodes using a wavelength demultiplexer.

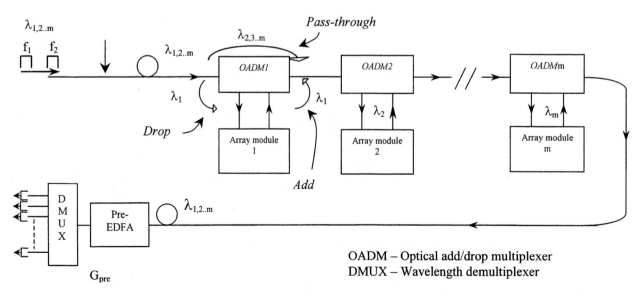

Figure 3: Time and wavelength division multiplexing architecture

In terms of optical power budget, this architecture is extremely efficient, and will allow for instance 8 array modules of 32 TDM hydrophones to be multiplexed onto a single telemetry fibre, giving a total of 256 hydrophone channels. The performance obtained from a single sensor multiplexed in this architecture has been investigated through a series of experimental and theoretical studies[8].

6. MULTIPLEXING EXPERIMENT – CROSSTALK MEASUREMENTS

In order to test the performance of the components required for the TDM/WDM architectures discussed above, a 3 wavelength version of the architecture shown in figure 2 was set up in the laboratory[7]. This system used a 3 wavelength Er doped DFB fibre laser with a wavelength spacing of 1.6 nm. Commercially available OADMs were used in the experiment. Crosstalk levels were measured by applying narrowband signals to individual array sensors using a piezoelectric fibre stretcher, and measuring the output of the adjacent sensor. Crosstalk between TDM channels within an array module was measured to be –47 dB. This level is believed to be set by the finite extinction ratio of the acousto-optic modulators. Crosstalk between sensors in adjacent modules (and therefore at adjacent optical wavelengths) was measured to be –76 dB, showing that dense WDM can be achieved with very low crosstalk between wavelengths.

7. PHASE RESOLUTION AND DYNAMIC RANGE OF MULTIPLEXED SYSTEMS

A theoretical analysis of the performance of the TDM/WDM architectures has been previously carried out[8], and the results are summarised here. This has been principally applied to hydrophone based systems, but also applies with minor modification to geophone arrays. The signal resolution can be determined by considering all the noise sources in the system. Since the signal of interest (acoustic or seismic) is converted into a phase modulation of the optical carrier, the bandwidth available for the signal to occupy will determine the theoretical maximum signal that can be measured. The ratio of the maximum signal to the signal resolution will then give the dynamic range.

7.1 Signal resolution
The signal resolution for a hydrophone system will be expressed as a minimum detectable pressure. For an interferometric system, this can be directly equated to a minimum detectable phase, taking into account the pressure /phase responsivity of the hydrophone (for a geophone, te relevant figure is the vibration./phase responsivity). The main sources of noise in an interferometric system will be:

a. Receiver noise
b. Shot noise
c. signal-spontaneous beat noise (if an optical preamplifier is used)
d. laser frequency noise
e. laser intensity noise
f. acousto-optic modulator drive signal noise (in a heterodyne system)
g. demodulator noise

The electronic system will be designed so that the demodulator noise is below other system noise sources, and with the most recent digital systems this is achievable in practice. In systems using an optical pre-amplifier, the receiver noise and shot noise will normally be dominated by the signal-spontaneous beat noise. If the hydrophone responsivity (i.e. the pressure/phase conversion) is given by r, then the pressure resolution due to the sensor self-noise δP_{self} is given by

$$\overline{\delta P_{self}^2} = \frac{1}{r^2} \cdot \left(\overline{\delta\phi_{min\_s-sp}^2} + \overline{\delta\phi_{laser-freq}^2} + \overline{\delta\phi_{laser-RIN}^2} + \overline{\delta\phi_{osc}^2} \right) \tag{1}$$

where $\delta\phi_{min\_s-sp}$ is the signal spontaneous beat noise, $\delta\phi_{laser-freq}$ is the laser frequency noise, $\delta\phi_{laser-RIN}$ is the laser RIN noise and $\delta\phi_{osc}$ is the oscillator noise. In practice, the minimum noise level may be determined by the ambient acoustic noise level of the environment in which the sensors are placed (the system will normally be designed to ensure this is the case). This can be added as an extra term to eqn 2.

7.2 Maximum signal

In the TDM reflectometric architecture with heterodyne detection, the received optical signal from a single sensor consists of a pulse of modulated intermediate frequency(IF) of frequency, $\Delta f(= f_1 - f_2)$, and pulse width, w, as shown in fig. 1, which is repeated at a frequency, f_{rep}. The acoustic signal appears as a phase modulation of the IF. The bandwidth available for the phase modulated signal to occupy can be shown to be equal to f_{rep}, where the value of f_{rep} is dependent on the number of sensors in a time multiplexed stream, N and the length of fibre per sensor, l. It can be shown using Carson's rule that the maximum signal, ϕ_{max} that can be accommodated in the available bandwidth is given by [3],

$$\phi_{max} = \left(\frac{c}{4lnNf_m} - 1 \right) \qquad (2)$$

where f_m is the signal frequency, n is the group refractive index of the fibre core and c is the speed of light in a vacuum. As the number of time-multiplexed sensors (or the fibre length/sensor) increases, f_{rep} decreases and so the allowable maximum signal reduces.

7.3 Sensor pressure resolution and dynamic range

The sensor pressure resolution and bandwidth limited dynamic range can be estimated in the two cases of self-noise limited and ambient acoustic noise limited using 2. The performance of the system architecture shown in figure 3 can then be analysed for the example case of 8 wavelengths and 32 hydrophones/wavelength (256 hydrophones/fibre pair total). Assuming that there is an optical power of −30 dBm returned from a single reflector to the pre-amp input, using typical amplifier performance values and normalising to a 1 Hz bandwidth then $\delta\phi_{min\_s-sp} = 13 \mu rad / \sqrt{Hz}$. The minimum and maximum detectable pressures for the two cases are plotted in fig 4a for the case of an in-water array with a hydrophone responsivity of −10 dB re 1 rad/Pa. The background noise floor is taken to be that found in deep water in the quietest conditions (known as Deep Sea State Zero). It can be seen that the calculated system self noise is well below the ambient noise at the frequencies of interest. The dynamic range of each sensor in the two cases are plotted in fig. 4b. We have taken the laser RIN to be −120 dB/Hz since this is typical for an optimised fibre laser source.

In the case of the sensor self-noise limited performance, the pressure resolution is limited by laser frequency noise at frequencies below 1 kHz and by pre-amp noise above 1 kHz. The dynamic range in this case is around 177 dB at 1 kHz and falls to around 110 dB at 10 kHz. When the hydrophone is placed in the ocean, the pressure resolution is limited by ambient acoustics below 6 kHz and pre-amp noise above 6 kHz. The effective dynamic range is now reduced to between 90 dB to 120 dB over the whole frequency range.

The analysis shows that for the in-water hydrophone case, the performance requirement specified in Table 1 can be achieved for a 256 channel multiplexed array. A similar analysis can also be carried out for a geophone based system.

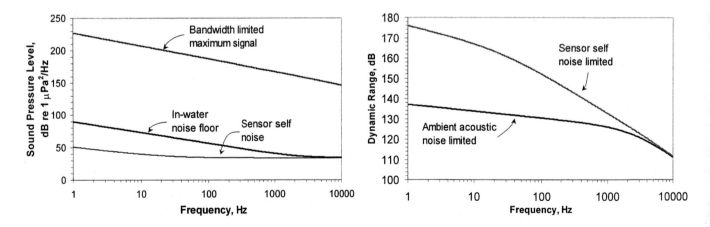

Figure 4(a): Minimum and maximum detectable pressures   (b) Sensor dynamic range

8. HYDROPHONE AND GEOPHONE DESIGNS

This paper has mainly concentrated on multiplexing and overall system design issues. We now discuss the important aspects of the sensor design that must be considered. A considerable amount of work has been carried out on both hydrophone and geophone designs at DERA and elsewhere, and this is summarised here.

8.1 Hydrophones

Fibre optic hydrophones have been very well developed for military applications, and their development has been widely reported in the literature[9]. One of the principal hydrophone performance factors is the pressure responsivity, expressed in radians/Pascal, which is a measure of the hydrophone efficiency in converting pressure into optical phase change. The required hydrophone responsivity will depend on the required system noise floor (normally expressed in Pa/√Hz, see Table 1) and on the achieved phase resolution of the interrogation system (in rad/√Hz), see eqn. 1. This responsivity must be maintained up to the maximum operating depth of the hydrophone (typically 500-1000m for military systems, implying a pressure level of 750-1500 psi). The hydrophone is also required to achieve a flat frequency response over the operating frequency range of interest (usually up to several kHz). This requires a device with the lowest resonance above the operating frequency band. The resonance will be dependent on the physical dimensions of the hydrophone, and the material properties of the fibre and encapsulant.

The requirements for streamer hydrophones are broadly similar to those for military systems. However, for deep-water OBCs, the hydrophone may need to operate down to a depth of 5000m, which requires a maximum operating pressure of 7500 psi. For the downwell hydrophone, operating pressure of up to 20000 psi may be required, together with temperature survivability up to 200 degrees C. These are demanding requirements which require careful material selection and hydrophone design.

Sonar hydrophones have typically been based on the air-backed mandrel design, on which the fibre coil is wound onto a length of tube, which is air-backed to increase the hydrophone response[10]. Hydrophones based on these designs typically achieve a normalised responsivity in the region of 0 to -10 dB re 1 rad/Pa over a frequency range of several kHz, using a fibre length of 20 – 100m. The system modelling described in section 7 assumed a hydrophone responsivity of –10 dB re 1 rad/Pa, and showed that with this responsivity the requires system signal resolution could be achieved for large multiplexed arrays. However, for downwell applications this responsivity must be maintained up to the maximum operating pressure of 20 000 psi. This is very difficult to achieve using an air-backed mandrel design. The responsivity is a function of the fibre length and the intrinsic pressure/phase conversion efficiency of the hydrophone. This efficiency can be increased by the use of an air backed mandrel, but will be limited by the requirement that for maximum operating depth, the strain induced in the fibre by the static pressure should be less than the elastic limit of the fibre (typically around 1%). To achieve the required responsivity, a longer length of fibre may therefore be needed, but this can reduce the system dynamic range, as can be seen from eqn. 2. For instance, to achieve a responsivity of –10 dB re rad/Pa at 20 000 psi maximum operating pressure(which is the most extreme requirement) it can be shown that around 1 km of fibre will be required in the hydrophone. Use of this fibre length would have implications for the system dynamic range, and would pose a significant packaging problem.

8.2 Geophones

Compared with hydrophones, fibre optic geophones are less well developed, although a number of devices have been reported. Geophones are basically vibration sensors and are typically based on a mass-spring system within a rigid case, in which the relative motion between the case and the mass is measured. In a fibre optic interferometric geophone, the relative displacement between the case and the mass in normally converted into a linear strain in a fibre coil. Such a device can be used to directly measure either displacement or acceleration, depending on whether the device is used above or below the fundamental resonance of the mass-spring system.

The key requirements for a practical geophone include high on-axis responsivity (sufficient to achieve the signal resolution stated in table 1), low cross-axis responsivity and small size (typically less than 30mm diameter, with ideally an aspect ratio of 1:1). The sensor should also have low responsivity to pressure signals and a frequency response which is flat over the frequency range of interest. High pressure tolerance can be achieved by making the geophone casing pressure resistant, but for downwell applications the sensor must also withstand the required temperature range, which will require appropriate material selection.

A number of fibre optic geophones have been reported. These include designs based on compliant mandrels and others based on mass-loaded flexural disks[11, 12]. Most of the devices reported have resonant frequencies greater than 300 Hz, and therefore operate below resonance as accelerometers. For both mandrel and flexural disk designs, the responsivity can be increased by increasing the size of the seismic mass, but this will also reduce the sensor bandwidth.

DERA have been developing a range of fibre optic accelerometers based on both mandrels and flexural disks for both military and other applications. The use of low-mass flexural disks as broadband accelerometers has been previously reported[13]. These devices have a frequency response which is flat up to 5 kHz and an acceleration responsivity of 17 dB re 1 rad/g. DERA have also developed a mass-loaded geophone optimised for low frequency operation. This device , shown in figure 5, achieved an on-axis responsivity of 32 dB re 1 rad/g with a frequency response flat to greater than 800 Hz. Although designed for low temperature operation, this geophone will be suitable with some modification for downwell operation.

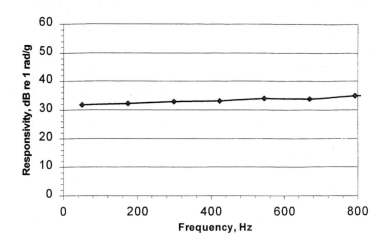

Figure 5: a. DERA geophone b. Geophone on-axis frequency response

9. DEVELOPMENT REQUIRED FOR GEOPHYSICAL APPLICATIONS

We have shown that fibre optic hydrophone technologies are already well developed for military applications, and that much of this technology can be transitioned readily into oilfield applications. While there is considerable overlap between the requirements of each application, a number of technical aspects need to be addressed before fibre optic hydrophones and geophones will be considered acceptable for practical systems. The more important aspects are considered to be:

- Ability of the sensor to handle large transient signals with low distortion. This will require optimisation of the demodulator performance. Most modern demodulator designs are based on digital approaches, which can be optimised for the particular signals of interest
- Achievement of low (<-30 dB) cross-axis sensitivity for geophones. This will be achieved through mechanical design of the sensor and possibly through the use of a twin fibre coil design
- Achievement of very low frequency signal resolution for geophones. This can be achieved through a combination of sensor design, and reduction of low frequency system noise
- Achievement of required responsivity with high static pressure tolerance (to 20 000 psi) for hydrophones. This can be achieved through the use of long fibre lengths, but reduction of system noise levels will also ease the requirement on hydrophone responsivity
- Proof of long term survivability at 150 to 200 ^{0}C for downwell array components, which will require extensive testing a and materials characterisation
- Engineering of arrays to survive deployment in the various environments.

A major concern is survivability in the downwell environments. However, many of the problems associated with this environment have already been overcome during the development of the existing static sensors.

10. CONCLUSIONS

It is now becoming accepted in the oil industry that the first generation of downwell fibre optic sensors have "come of age". These devices, designed for the measurement of quasi-static quantities such as temperature and pressure, are now being installed in significant quantities in oilfields around the world. Interest is now growing in a new generation of sensors based on the measurement of dynamic quantities, in particular for seismic measurements in environments ranging from oil wells to the ocean floor. Fibre optic sensors have many advantages for these applications, but the significantly different requirements pose a new set of problems to be overcome. Many of these issues have already been addressed in the development of hydrophone systems for military applications. In particular, an approach being developed by DERA based on time and wavelength division multiplexing of interferometric sensors offers many advantages for seismic applications.

REFERENCES

1. A. Mendez, R. Dalziel and N. Douglas, "Applications of optical fiber sensors in subsea and downhole oil well environments", *Harsh Environment Sensors II*, SPIE vol. **3852**, pp 16-28, 1999
2. A. Kersey, "Optical fiber sensors for downwell monitoring applications in the oil and gas industry, *Optical Fiber Sensors 13*, SPIE vol. **3746**, pp326-331, 1999
3. R. G. May et al, "SCIIB pressure sensors for oil extraction applications", *Harsh Environment Sensors II*, SPIE vol. **3852**, pp 29-35, 1999
4. P. Nash, "Review of interferometric optical fibre hydrophone technology", *IEE Proc. Radar, Sonar and Navigation*, **143, No. 3**, pp 204-209, 1996
5. A. Dandridge, "Development of fiber optic sensor systems, *Optical Fiber Sensors 12*, SPIE vol. **2360**, pp 154-161, 1994
6. J. Dakin, "Novel optical fibre hydrophone array using a single laser source and detector", *Electronic Letters,* vol. **20**, pp. 53-54, 1984
7. G.A. Cranch and P.J. Nash, "High multiplexing gain using TDM and WDM in interferometric sensor arrays, *Fibre Optic Technology and Applications*, SPIE vol. **3860**, pp. 531-537, 1999
8. G.A. Cranch and P.J. Nash, "Large-scale multiplexing of interferometric sensors using TDM and DWDM with low crosstalk", to be submitted for publication in the J. Lightwave Technology.
9. J. Bucaro, N. Lagakos, J. Cole and T. Giallorenzi, "Fibre optic acoustic transduction, *Physical Acoustics* **XVI**, pp. 385-457, 1982
10. S knudsen, A. B. Tveten, A. Dandridge and K. Blotekjaer, "Low frequency transduction mechanisms of fiber-optic air-backed mandrel hydrophones", Optical fiber Sensors 11, pp208-211, 1996
11. D.L. Gardner and S.L. Garrett, "Fiber optic seismic sensor", *Fiber Optic and Laser Sensors V*, SPIE vol. **838**, pp. 271-278, 1987
12. D.A. Brown and S.L. Garrett, "An interferometric fiber optic accelerometer", *Fiber Optic and Laser Sensors IV*, SPIE vol. **1367**, pp. 282-299,
13. G. Cranch and P. Nash, High Responsivity Fibre Optic Flexural Disk Accelerometers, Accepted for publication in Journal of Lightwave Technology

Optical detection of gas in producing oil wells

F. Cens, B. Théron, D. Vuhoang, M. Faur, F. Rezgui, D. McKeon

Schlumberger Riboud Product Center, Clamart, France

ABSTRACT

A new production logging device has been field tested that uses innovative sensing technology to enable the direct detection and quantification of gas in multiphase flows. Four optical probes, deployed 90 degrees apart on the arms of a centralizer-like tool, measure the optical reflectance of the surrounding fluid. The probes are evenly spaced in the pipe cross section, and their orientation in space is accurately known through use of an integrated relative-bearing sensor. In gas-liquid mixtures, the optical signal reflected by the probe is used to determine gas holdup and a gas bubble count, which is related to gas flow rate.

In addition, the individual sensor measurements are used to build an image of the gas flow in the well. These images are particularly useful in deviated and horizontal wells for better understanding the multiphase flow patterns and interpreting their inherent phase segregation occurring at such deviations.

The new tool has been successfully field tested in wells throughout the world and the tool's capabilities are illustrated by example from both field and laboratory data sets.

The new tool has been designed to detect the presence of gas, and hence its major application is to identify gas entries in oil/water wells or water/oil/condensate in gas wells. Because of its high sensitivity to minute amounts of gas, the tool can also be used to locate the bubble point when logging in the tubing.

The introduction of optical sensing technology in this new tool represents an innovation in production logging. The provided data enable the direct detection and quantification of gas or liquid in multiphase mixtures, allowing the precise diagnosis of well problems and helping in design of production enhancement interventions.

Keywords: Gas detection, optical sensor, well, multiphase flow, hold-up, bubble count, entries detection

1. INTRODUCTION

The Gas Holdup Optical Sensing Tool (GHOST[*]) has been developed using innovative optical sensing technology to enable the direct detection of gas in producing wells. In gas-liquid mixtures, the optical signals reflected by the new optical probes vary from liquid to gas values in a binary manner. This allows the determination of the gas holdup and the gas bubble count. The GHOST tool is deployed as a new module in the Schlumberger existing production logging string. Field tests throughout the world have demonstrated the tool's capabilities to detect gas entries and to quantify gas holdup.

---

[*] Mark of Schlumberger.

2. PRINCIPLES

2.1. GHOST MEASUREMENT PRINCIPLE

Gas detection is achieved using an optical sensor sensitive to the optical index of the fluid. The optical index of a gas is close to 1, for water it is 1.35 and for crude oils it is typically 1.5. Figure 1 shows the schematic of the optical system. Light is emitted by an LED into a fiber, and travels up to the sapphire-probe tip. Then, depending on the optical index of the fluid, some light is reflected back to the photodiode, which converts the light to an electrical signal. The amount of reflected light is large when the fluid is gas, but the signal amplitude is low when the tip is in a liquid. The system is somewhat binary with a large difference between gas and liquid signal amplitudes. A threshold is set in order to differentiate gas from liquid (signal > threshold ➔ gas).

From the raw output signal waveform , the local gas holdup is computed as follow (Figure 2):

$$Y_g = \frac{t_g}{t_g + t_l}$$

where t_g and t_l are the total time the probe tip spent in gas and liquid, respectively.

Another important quantity can be quantified from the raw data, namely, the gas bubble count. The rate of gas bubbles passing a probe is determined by dividing the number of times that the signal crosses the threshold line per second by two.

$$B_c = \frac{n_{crossing}}{2(t_g + t_l)}$$

This reflects the number of gas or liquid bubbles or drops detected per second by the probe.

Furthermore, [ref. 1], field tests have shown that under favorable conditions the GHOST measurement can also differentiate oil from water. Figure 3 shows the amount of light reflected as a function of the fluid optical index. In liquid, two sub-levels can be defined: the lowest level accounts for oil and the upper for water; an example of a raw signal is presented in Figure 4. For the time being, this method is still qualitative, and further research is underway to develop a quantitative answer.

2.2 TOOL OVERVIEW

Four optical probes are deployed on four arms to obtain good coverage of the well cross section area. A schematic of the tool is provided on Figure 5. A caliper and a relative bearing sensor included in the tool are used to determine the exact position of the optical probes in the well cross section.

It is corrosion resistant and is compliant with NACE requirements.

2.3. LABORATORY TESTS

The GHOST has been extensively tested in a laboratory flowloop over a large range of gas-liquid flowrates. Gas holdup from the GHOST has been compared with independent holdup measurements (squares on Figure 6). The flowloop tests show that for high and low gas holdup the measurements are in good agreement with the reference holdup; the difference is greater for a 50% gas mixture. An empirical approach, based on laboratory data, has been derived to correct the measurements (solid line on Figure 6) and has been validated against field data.

2.4 INTERPRETATION THEORY

The holdups of the different fluids gas, oil and water (Y_g, Y_o and Y_w) are derived solving the following equation:

$$Y_w + Y_g + Y_o = 1 \qquad \text{(eq. 1)}$$

Optical (Y_g) and electrical data (Y_w) are used to solve for Y_o (eq. 1)

Then the different flowrates, Q_w, Q_o and Q_g, are computed using slippage velocities (V_{so}, V_{sg}) according to:

$$Q_w = Q_t\,Y_w - Q_{so}/(Y_w+Y_o) - Q_{sg}\,Y_w/(Y_w+Y_o) \qquad \text{(eq. 2)}$$
$$Q_o = Q_l - Q_w \qquad \text{(eq. 3)}$$
$$Q_g = Q_t - Q_l \qquad \text{(eq. 4)}$$

where A is the well cross section area, Q_t the total well flowrate (derived from a sonde able to measure the mean velocity of the fluid mixture and the diameter of the pipe):

$$Q_{so} = Y_w\,Y_o\,V_{so}\,A \qquad \text{(eq. 5)}$$
$$Q_{sg} = (Y_w+Y_o)\,Y_g\,V_{sg}\,A \qquad \text{(eq. 6)}$$
$$Q_l = (Y_w+Y_o)\,Q_t - Q_{sg} \qquad \text{(eq. 7)}$$

3. FIELD EXAMPLE

The well is an oil-gas well drilled with a 5 degrees deviation, completed with a 7"casing and with a surface production rate of 7700 BOPD, 4000 BWPD and 48 MMSCFD of gas. A GHOST and FLOWIEW combination was run to determine the contribution of each producing zone. The FLOVIEW tool has been developed to enable the direct water detection. It is equipped with four electrical probes, which are sensitive to the conductivity of the fluid present on its tip. Typically the impedance "seen" by the probe varies strongly when it is in water or in hydrocarbon. It also measures the tubing diameter thanks to a caliper and the velocity of the fluid mixture thanks to a spinner.

The GHOST and FloView measurements detected stagnant water at the bottom of the well, whereas the gradiomanometer (tool measuring the fluid mixture density) located 19.5 ft above had its first reading already in the oil producing zone.

First oil and gas entries from the lowest set of perforations would have gone unnoticed by the gradiomanometer. The GHOST and FloView holdup and bubble count images/curves clearly detect these entries (Figure 7) at xx121 ft, a bare 8 ft from reachable total depth. The low gas holdup recorded over this lower set of perforations suggested that most of the produced fluid is oil. Higher up, over the second set of perforations, water entry is observed at xx080 ft, where the FloView water holdup starts increasing until xx065 ft. At this point, gas begins entering the borehole with the main gas entry at xx028 ft☐ Due to the GHOST tool's finer sensing zones that are ten times smaller than that of the FloView tool (100 ☐m vs. 1 mm), the GHOST probes detect more bubbles than the FloView probes. Comparison of the GHOST bubble count images to the FloView bubble count images confirms this (Figure 7) since both images use the same color scale that goes from light grey (1 bubble) to black (2000 bubbles).

4. CONCLUSION

The paper describes a new production logging device that uses innovative sensing technology to enable the direct detection and quantification of gas in multiphase flows. Four optical probes, deployed 90 degrees apart on the arms of a centralizer-like tool, measure the optical reflectance of the surrounding fluid to discriminate gas from liquid. The new tool provides gas holdup, gas bubble count, and a flow profile image. The latter is particularly useful in deviated and horizontal wells for better understanding and interpretation of the inherent phase segregation occurring at such deviations.

Major applications of the GHOST tool include:

- identification of gas entries in oil/water wells

- identification of liquid entries in gas wells

- detection of small oil entries in three-phase flows

- stand-alone tri-phasic evaluation (under favorable conditions).

Unlike the gradiomanometer tool, the new GHOST tool is insensitive to friction effects. In addition, the GHOST probes are sensitive to small amounts of gas, and do not require a downhole calibration. Furthermore, when combined with an electrical probe tool measuring water holdup, a complete, unique three-phase flow diagnosis can be made.

5. ACKNOWLEDGEMENTS

The authors would like to thank all the operating companies that helped in field testing the new logging tool and particular BP-Amoco management for their permission to publish the data presented here. Thanks to J. Flores who reviewed this paper. We also share the success of the tool with all the Schlumberger personnel who developed and field tested the GHOST tool.

6. REFERENCES

ref. 1 Patent N°96.06361 French application of May 22nd, 1996 ref. *2* J. G. Flores et al., 2000, *"Determining Intervals of lean Gas Breakthrough in a Gas Condensate Producer Using an Optical Holdup Meter"* ASME ETCE/OMAE Joint Conference, New Orleans, Louisiana

Figures:

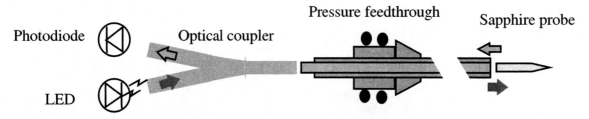

Figure 1 Opto electronic chain schematic

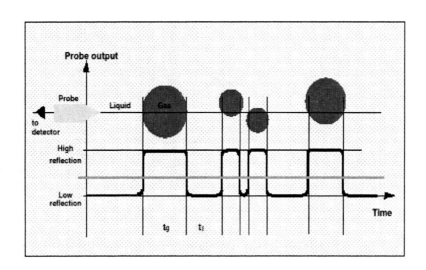

Figure 2 Theoretical raw signal (waveform) from optical probes

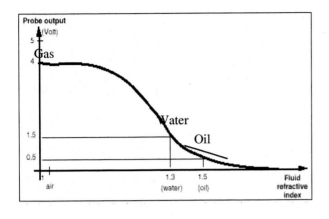

Figure 3 Optical Probe Response

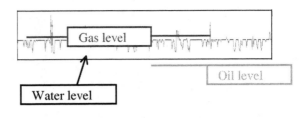

Figure 4 Raw optical signal in a Water/Oil/Gas mixture

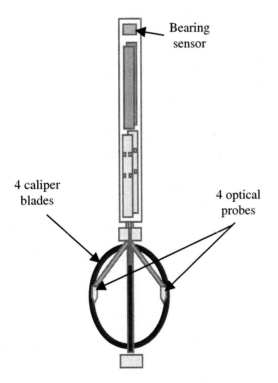

Figure 5 Schematic of the GHOST tool

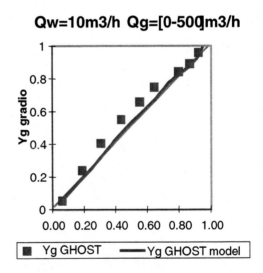

Figure 6 GHOST Gas Holdup comparison with the reference gradiomanometer and model

Figure 7 Log and flowrate computation in K5 well

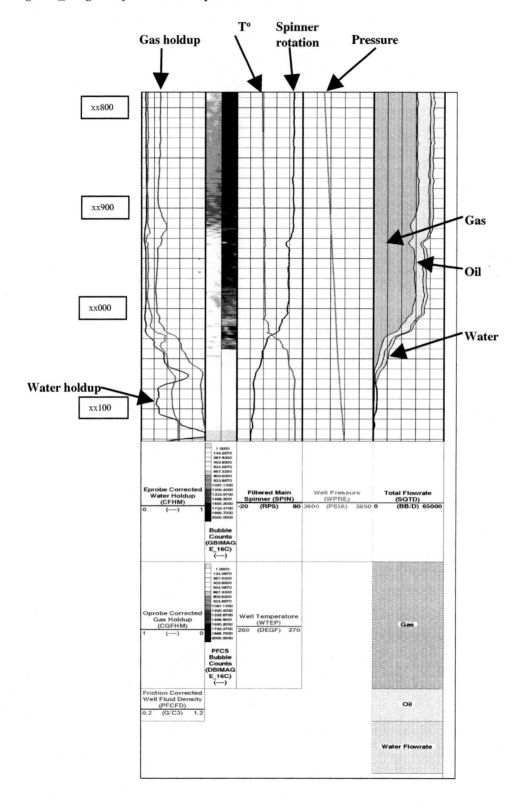

Continuous-Phase Fluid Velocity Measurement for Oil Field Flows Using Local Optical Probe and Tracing Method

Rogerio T. Ramos and Andrew Holmes*

ABSTRACT

The measurement of the velocities of oil and water phases in the bore-hole of a producing well, together with volume fraction measurement, gives the flow rate of each fluid phase. Even though radioactive tracers methods provides a good velocity measurement, safety concerns and regulations makes their use unattractive.

We propose a method of measuring fluid velocity using local optical fiber probes and tracer techniques. A fluorescence dye is injected into the bore-hole where it is mixed with the fluid of interest. An optical fiber probe is positioned down-stream from the point of injection detects the passage of the tracer with the flow. The time between injection and detection enables one to calculate the flow velocity for a given distance between injector and detector. The same local probes could also be used for hold-up estimation.

We describe laboratory experiments using a flow loop, fluorescence dye as tracer and an optical fiber probe. The dependency of the injection to detection distance to the flow mixing is observed. Best results for average velocity estimation can be achieved for long distances. Measurements at short distances are possible but some kind of flow model may be needed to interpret the results.

Keywords: O ptical fiber sensors, fluorescence, multi-phase flow

1. INTRODUCTION

The measurement of the velocities of oil and water phases in the bore-hole of a producing well, together with volume fraction measurement, gives the flow rate of each fluid phase. This data can be provided by a tool connected to the surface by a wire-line cable and used to survey the well at different depths.

Radioactive tracers have been used to obtain fluid flow velocity.[1] The time between injection and detection enables one to calculate the flow velocity for a given distance d between injector and detector. Even though this method provides a good velocity measurement, safety concerns and regulations makes the use of radioactive tracers unattractive. Furthermore, a different kind of measurement is needed for the measurement of the volume fraction of each phase in order to estimate individual flow rates.

The use of fluorescence tracers also seems attractive due to high sensitivity even at low concentrations. But what makes it even more desirable is the possibility of using the same detection system used for volume fraction measurement, without the need of a pulsed neutron generator. This article explores the possibility of using local optical probes to measure fluid phase velocities with fluorescence tracer injection.

2. LOCAL PROBES

Local electrical probes have been used for volume fraction measurements. One specific tool, uses an array of local conductivity probes to detect the water content in the flow.[2] Local optical fiber probes have been used as the basis for multiphase flow metering in air-water or steam-water flows in the power industry. Local fiber sensors were consider for petroleum engineering contexts, and for associated studies of model oil/water/gas flows in laboratory flow loops. The first essential of such probes is the ability to discriminate drops, bubbles or other moieties of the three immiscible phases. The simplest optical measurements to make are of reflectivity (reflected power ratio) at the end of an optical fiber.[3] This kind of measurement has proved to provide good estimation of the volume fraction of each phase.[4][5][6]

The same local probe can be made sensitive to an appropriate tracer injected up-stream. An array of local probes could then provide a map of where the tracer is being sent in the pipe. A time of flight measurement can provide fluid velocity. The tracer can be made soluble into the fluid phase of interest.

Rogerio Ramos is with Schlumberger-Doll Research, Old Quarry Road, Ridgefield, CT 06877-4108, U.S.A., and Andrew Holmes is with Schlumberger Cambridge Research, High Cross, Madingley Road, Cambridge CB3 0EL, U.K., where this work was done. E-mail: ramos@ridgefield.sdr.slb.com

3. FLOW MIXING

In general, there are three cases to be considered when using local probes to detect tracing material:

1. $d \ll D$

2. $d \approx D$

3. $d \gg D$

Where D is the pipe diameter and d is the distance between injection and detection. In the case where $d \ll D$, it can be considered that the measured velocity corresponds to the local velocity of the fluid. Normally the Reynolds number for flows expected down-hole would be Re $\gg$ 2000, which means the the flow can often be considered turbulent.

When $d \approx D$, the tracer will be partially mixed to the fluid in the flow when it arrives at the detector.[9] The signal corresponding to the concentration of tracer would not form a continuous distribution with time, but it would tend to present a series of peaks corresponding the pockets of tracer. In this case, the time of arrival of the first peak could be related to the maximum velocity of the fluid velocity distribution in the pipe for the position of the detection probe in the pipe. As a first approximation for a turbulent flow, this would happen for $D < d < 100D$.

For $d \gg D$, the tracer should be well mixed to the fluid in the flow and the signal corresponding to the concentration of tracer would form a continuous distribution with time. The time of arrival of the peak of this distribution can then be linked to the average velocity of the flow.[9] As a first approximation for a turbulent flow, this would happen for $d > 100D$.

It is important to take these three cases in mind when thinking of making a phase velocity measurement using tracer techniques over short lengths, as the kind of velocity measured can be different.

4. EXPERIMENTS

Experiments were made using a single optical fibre probe and the injection of a fluorescence tracer (fluorecin KeyTec yellow tracer R). Volumes of about 1 ml of tracer were injected for each measurement. Figure 1 gives a diagram of the experimental arrangement. The light source used was a green HeNe laser, 544 nm, 800 μW. A 140 μm fibre with a 100 μm core was used in this test. A filter with a cut off wavelength of 580 nm was mounted between collimating lenses. At least 60% of the power was lost in the collimating optics.

The tests were made for a single phase horizontal flow of kerosene, with dynamic viscosity $\mu \approx 0.001$ N s m^{-2} and mass density $\rho \approx 800$ kg m^{-3}. A pipe diameter $D = 0.05$m and fluid velocities $v \geq 0.5$ m s^{-1} (up to 1.5 m s^{-1}) give Reynolds numbers Re$\geq$ 25000, (Re $= \frac{vD\rho}{\mu}$). Two data sets were collected for two cases: $d = 10D$ and $d = 100D$.

4.1. $d = 10D$ experiments

Figure 2 gives an example of the fluorescence signal detected. The presence of multiple peaks indicates that there are points of higher concentration of tracer than others. In this case, the tracer is not well mixed to the fluid flow at the time of the measurement and the first arrival method is used to calculate the maximum velocity. Figure 3 shows the measured maximum velocity plotted against the reference average velocity given by a turbine meter.

The average velocity can be estimated from the maximum velocity by using a model of the flow velocity distribution. The velocity distribution inside the pipe with a turbulent flow can be approximated by[10]:

$$u \approx u_{max}(\frac{R - r}{R})^{1/7} \tag{1}$$

Where u the local velocity at a radial position r, u_{max} is the maximum velocity and R is the pipe radius.

The average velocity $\bar{u}$ would be given by:

$$\bar{u} = u_{max}\frac{2\pi}{\pi R^2}\int_0^R (\frac{R - r}{R})^{1/7}rdr \tag{2}$$

Figure 4 shows the estimated average velocity from the measured maximum velocity plotted against the reference average velocity given by a turbine meter. Good agreement with the reference has been achieved in this case (errors within $\approx$ 10%).

4.2. $d = 100D$ experiments

Figure 5 gives an example of the fluorescence signal detected. The presence of a single peak indicates that there is a good mix of the tracer to the fluid flow and the time between injection and this peak can provide an average velocity. Figure 6 shows the measured average velocity plotted against the reference average velocity given by a turbine meter with good agreement (errors well within 10%). The velocities were calculated by using the time between injection and the time when the detected fluorescence signal is maximum.

5. COMMENTS

The dependency of the injection to detection distance to the flow mixing was demonstrated. Best results for average velocity estimation can be achieved for long distances ($d \gg 100D$). Measurements at short distances are possible but some kind of flow model may be needed to interpret the results.

Signal filtering and other signal processing techniques could be used to better estimate the phase velocity. Combining the signals from an array of probes in an appropriate way would provide a better estimation of the average phase velocity and provide information about the velocity distribution of the continuous phase.

The main advantage of the system proposed here is the possibility of employing the same probe, or array of probes, used for volume fraction estimation to perform the continuous phase velocity measurement. This can lead to shorter tool strings. A smaller quantity of fluorescence tracer per shot required could increase the number of shots. This tracer is also potentially less aggressive than the Gadolinium tracer used today.

Acknowledgments

We thank S. Coupland and G. Thomas for experimental data collection, and K. Stephenson, John Ferguson, O. Mullins and E. Fordham for the many discussions.

REFERENCES

1. A.D. Hill & J.R. Solares. "Imporved analysis methods for radioactive tracer injection logging". *Journal of Petroleum Technology*, pp. 511-520, March, 1985.

2. F.R. Halford, S. Mackay, S. Barnett & J.S. Petler. "A production logging measurement of distributed local phase holdap". *Society of Petroleum Engineers*, SPE 35556, 1996.

3. R.T. Ramos, E.J. Fordham, "Oblique-tip Fiber-optic Sensors for Multi-phase Fluid Discrimination", *Journal of Lightwave Technology*, vol. 17, n. 8, pp. 1392-1400, August, 1999.

4. E.J. Fordham, A. Holmes, R.T. Ramos, S. Simonian, S.-M. Huang & C.P. Lenn. "Multi-phase fluid-discrimination with local fibre-optical probes: I. Liquid/liquid flows". *Measurement Science & Technology*, vol.10, n.12, pp.1329-1337, December 1999.

5. E.J. Fordham, S. Simonian, R.T. Ramos, A. Holmes, S.-M. Huang & C.P. Lenn. "Multi-phase-fluid discrimination with local fibre-optical probes: II. Gas/liquid flows". *Measurement Science & Technology*, vol.10, n.12, pp.1338-1346, December 1999.

6. E.J. Fordham, R.T. Ramos, A. Holmes, S. Simonian, S.-M. Huang & C.P. Lenn. "Multi-phase-fluid discrimination with local fibre-optical probes: III. Three-phase flows". *Measurement Science & Technology*, vol.10, n.12, pp.1347-1352, December 1999.

7. X. Wu, E.J. Fordham, O.C. Mullins, R.T. Ramos. "Single Point Optical Probe for Measuring Three-Phase Characteristics of Fluid Flow in a Hydrocarbon Well". *US patent* 6,023,340, granted on February 8, 2000.

8. R.T. Ramos, K.E. Stephenson. "Apparatus and Tool using Tracers and Single Point Optical Probes for Measuring Characteristics of Fluid Flow in a Hydrocarbon Well and Methods of Processing Resulting Signals". *US patent* 6,016,191, granted on January 18, 2000.

9. G.I. Taylor. The Dispersion of Matter in Turbulent Flow Through a Pipe. *Proceedings of the Royal Society*, A, vol.CCXXIII, 446-68, 1954.

10. G.W. Govier & K. Aziz. The Flow of Complex Mixtures in Pipes. Van Nostrand Reinhold, New York, 1977.

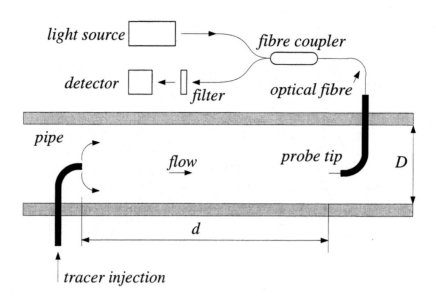

Figure 1: Diagram of the experimental arrangement used in the tests.

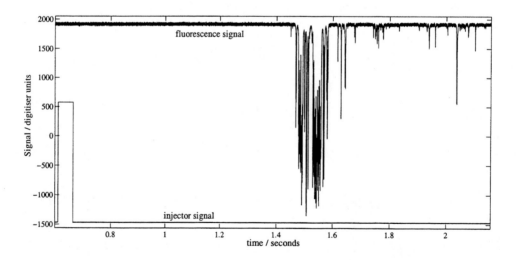

Figure 2: An example of the fluorescence signal detected for the $d = 10D$ experiments. The square pulse at the bottom of the plot indicates the time of injection. (2048 digitiser units = 5V)

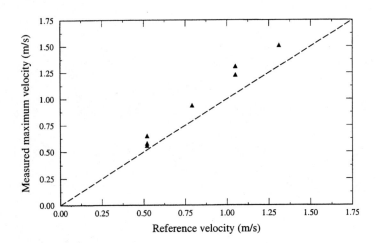

Figure 3: Measured maximum velocity plotted against the reference average velocity given by a turbine meter (for $d = 10D$).

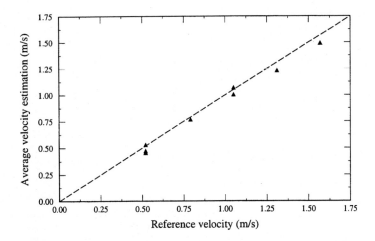

Figure 4: Estimated average velocity from the measured maximum velocity plotted against the reference average velocity given by a turbine meter (for $d = 10D$).

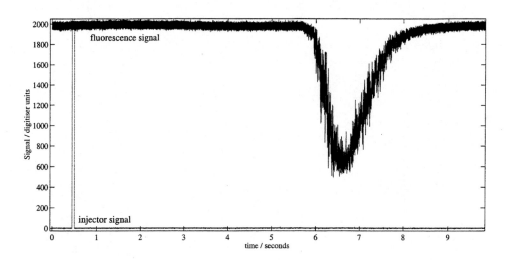

Figure 5: An example of the fluorescence signal detected for the $d = 100D$ experiments. The square pulse at the bottom of the plot indicates the time of injection. (2048 digitiser units = 5V)

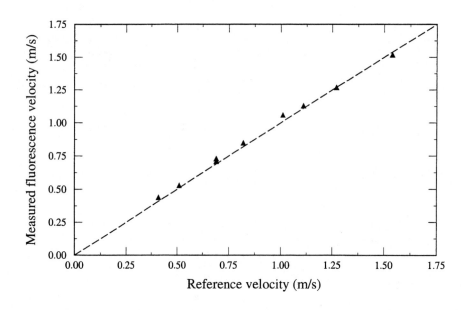

Figure 6: Measured average velocity plotted against the reference average velocity given by a turbine meter (for $d = 100D$).

Static and dynamic testing of bridges and highways using long-gage fiber Bragg grating based strain sensors

Whitten L. Schulz[*a], Joel P. Conte[b], Eric Udd[a], John M. Seim[a]

[a]Blue Road Research, 2555 NE 205th Ave., Fairview, OR 97024.

[b]Dept. of Civil and Env. Engrg., Univ. of California, Los Angeles

ABSTRACT

Fiber optic Bragg gratings packaged in long gage configurations are being used to measure static and dynamic strain in structures and structural models to monitor structural health and predict damage incurred from a seismic event. These long gage sensors are being used to experimentally verify analytical models of post-earthquake evaluation based on system identification analysis. This fiber optic deformation measurement system could play a significant role in monitoring/recording with a higher level of completeness the actual seismic response of structures and in non-destructive seismic damage assessment techniques based on dynamic signature analysis. This new sensor technology will enable field measurements of the response of real structures to real earthquakes with the same or higher level of detail/resolution as currently in structural testing under controlled laboratory conditions.

Keywords: Macroscopic strain, dynamic strain measurements, optical

1. INTRODUCTION

Determination of the actual nonlinear inelastic response mechanisms developed by civil structures such as buildings and bridges during strong earthquakes and post-earthquake damage assessment of these structures represent very difficult challenges for earthquake structural engineers. Presently, there is an unbalance between the analytical capabilities for predicting various nonlinear structural responses and damage parameters and the incompleteness (lack of richness) of the information on the actual seismic response of structures measured in the laboratory and, more importantly, in the field. Furthermore, this unbalance impedes the full deployment of the new and very appealing philosophy of performance-based earthquake engineering.

The research being performed aims at filling the gap defined above through studying the feasibility, through physical experimentation at small scale, of using long gage fiber optic Bragg grating sensors for monitoring directly the "macroscopic" internal deformation response of structures to strong ground motions and for non-destructive post-earthquake evaluation of structures. These fiber optic sensors can be either embedded inside a reinforced concrete or composite structure or bonded to the surface of a steel structure and monitored real-time at speeds in excess of 1kHz.

1.1. Fiber Bragg Grating Sensor

The core sensor technology for this project is the fiber Bragg grating. A Bragg grating consists of a series of perturbations in the index of refraction along the length of a fiber (Udd, 1991). This grating reflects a spectral peak based on the grating spacing, thus changes in the length of the fiber due to tension or compression will

---

* Correspondence: Email: whitten@teleport.com; WWW: http://www.bluerr.com; Telephone: 503 667 7772 x102; Fax: 503 667 7880

change the grating spacing and the wavelength of light that is reflected back. Quantitative strain measurements can be made by measuring the center wavelength of the reflected spectral peak. Fig. (1) shows a Bragg grating and the effects of a broadband light source impinging on the grating.

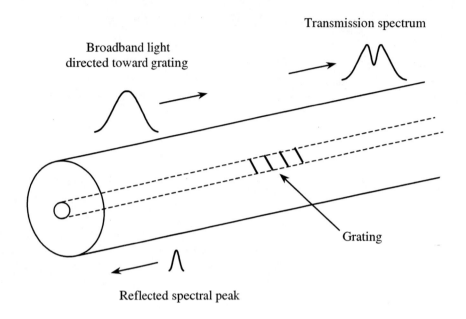

Fig. 1 Transmission and reflection spectra from fiber Bragg grating

In its basic form, a typical Bragg grating has a gage of approximately 5mm. For most civil structure applications this gage is too short, so a method of effectively increasing the gage was developed.

1.2. Long Gage Grating Sensors

In order to increase the gage of the Bragg grating to provide a more macroscopic strain value useful in civil structure applications, the grating is packaged in a tube with the tie points defining the effective gage. Fig. (2) shows a long gage sensor with optional brackets for surface mounting.

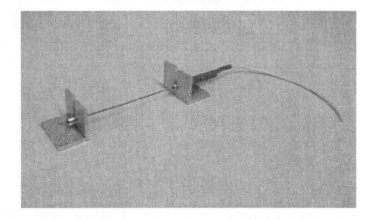

Fig. 2 Long gage grating sensor shown with optional brackets for surface mounting capability

This packaging provides a gage range from 2.5 to 100 cm. The maximum diameter of the grating package is less than 8mm, making it non-obtrusive and ideal for embedding into composites, placing into grooves in concrete, etc.

1.3. Dynamic Strain Measurement

The analytical models in this project require dynamic strain measurements for experimental verification. A high-speed fiber grating demodulation system has been developed that can measure strain from DC levels up to 10Mhz (Seim, 1998). For this project, the system has been optimized for 1 kHz, which provides sufficient oversampling. This demodulation system consists of a grating filter that converts the spectral information from the grating sensor into an amplitude based signal measurable by photo detectors. Fig. (3) shows this system where light is directed into the sensor through the first beam splitter, reflected in the grating and directed into the second beam splitter where it is divided into the filtered and reference legs and then into the high speed detector. The reference leg compensates for amplitude losses in the system.

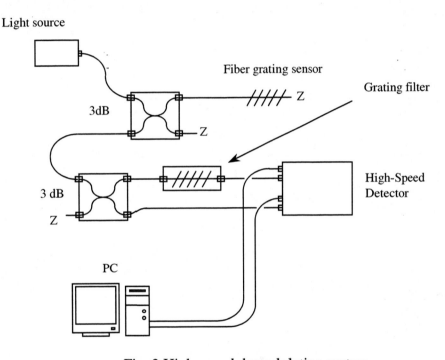

Fig. 3 High-speed demodulation system

This high-speed system can also be expanded to support more than one grating sensor as represented in Fig. (4).

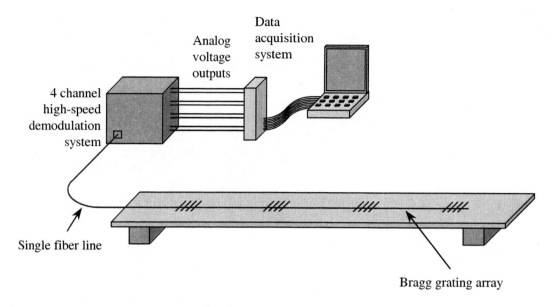

Fig. 4 High-speed measurement of a series of fiber gratings along a single fiber line

2. CIVIL STRUCUTRE EXAMPLES

Feasibility is an important aspect of introducing new technology and the sensors and system described above have been used on an actual structure with good success. One of the examples below has been providing long-term survivability and sensor performance data while a new project has the potential of monitoring a bridge during demodulation.

2.1. Horsetail Falls Bridge

Twenty-six sensors have been successfully monitoring the Horsetail Falls Bridge in Oregon for two years (Seim, 1999). The bridge, shown in Fig. (5), was built in 1914 and in 1998 underwent a strengthening procedure where composite wrap was placed over the concrete beams.

Fig. 5 Horsetail Falls Bridge before being strengthened by composite wrap and instrumented with 26 long gage fiber grating strain sensors

To verify that the composite wrap was adding strength to the bridge, long gage fiber grating strain sensors were placed in grooves cut into the concrete and in the wrap itself. Fig. (6) shows the sensors being embedded into the concrete and composite wrap.

Fig. 6 Sensors being placed into grooves in the concrete (left) and embedded into the composite wrap (right)

Using the high-speed demodulation system, several dynamic tests were performed on the bridge. Fig. 7 shows results from one of the tests focusing on one sensor as traffic was monitored.

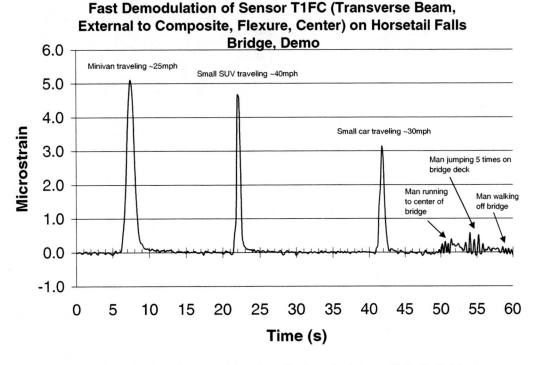

Fig. 7 Dynamic data from a sensor installed on the Horsetail Falls Bridge

The high-speed demodulation system has a high sensitivity as demonstrated by the resolution of less than 0.1 microstrain in the above data.

2.2. Sylvan Bridge

In mid July, 2000 14 long gage sensors were installed on another Oregon bridge shown in Fig. (8). This bridge also received composite wraps, but in strips as opposed to sheets. This could be an important part of this seismic damage assessment project as the bridge is tentatively scheduled to be destroyed in 3-5 years. This will allow for severe, damaging forces to be monitored with the sensors and compared to analytical models.

Fig. 8 Sylvan Bridge in Oregon scheduled for instrumentation mid July 2000

3. EXPERIMENTAL VERIFICATION OF THE ANALYTICAL SEISMIC DAMAGE MODEL

While the applications listed above provide excellent structural health and feasibility information, it is still necessary to use models to verify the analytical seismic damage model being developed. These experiments will be conducted in two main stages, free and forced vibration. Damage will also be induced during this dynamic testing and compared to predicted damage locations.

3.1 Free Vibration

The first set of experiments will employ an aluminum beam with various combinations of boundary conditions and mass distributions. The beam will be approximately 100 cm long and have three sensors with gages of 15 cm. Fig. 9 shows a representation of the beam and the various configurations. Several long gage sensors will be placed at key locations on the beam to measure macro strain during free vibration. These strain values will then be compared to results from the analytical model.

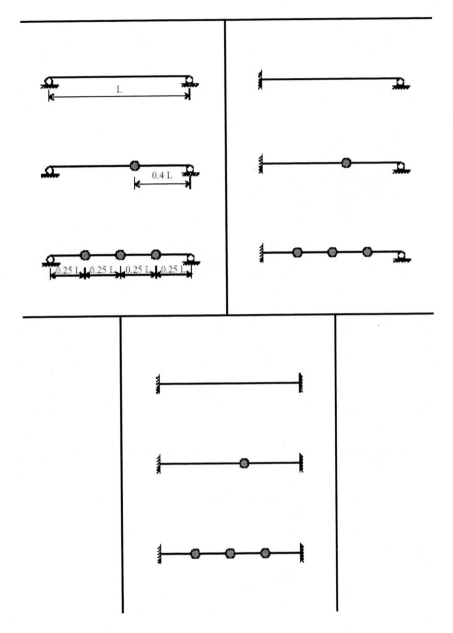

Fig. 9 Different mass distributions and boundary conditions for test beam in free vibration

This rigorous set of free vibration experiments will better verify the analytical model and the long gage sensors.

3.2 Forced Vibration

The second main set of experiments will involve forced vibrations with more complex models better representing real structures. Fig. 10 shows the forced vibration setup with a shaker table and examples of potential models.

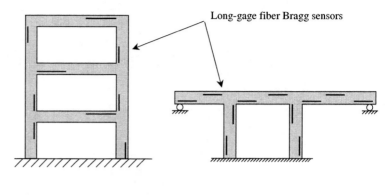

Fig. 10 Shaker table setup (left) and structural models (right) of forced vibration tests

3.3. Induced Damage

The analytical model has the capability to predict where damage has occurred based on dynamic signature analysis. During dynamic testing, damage will be simulated by reducing the cross sectional area of the models at known points to see if the analytical model gives the same damage location.

4. SUMMARY

Long gage fiber optic Bragg grating sensors are being used to monitor the structural health of structures and to provide experimental verification of a seismic damage identification model. The long gage is to provide a macroscopic strain value more useful in structural monitoring. These sensors have the capability of measuring dynamic strain up to several kHz, which is an important aspect of seismic monitoring and testing.

ACKNOWLEDGEMENTS

The work for this project has been partially funded by The National Science Foundation (Grant # DMI-9961268)and The United States Navy (Contract # N68335-99-C-0242.)

REFERENCES

1. Seim, J., E. Udd, W. Schulz, and H.M. Laylor (1999). "Composite Strengthening and Instrumentation of the Horsetail Falls Bridge with Long Gauge Length Fiber Bragg Grating Strain Sensors", *SPIE Proceedings*, Vol. 3746, p. 196, 1999.

2. Seim, J.M., W.L. Schulz, E. Udd, K. Corona-Bittick, J. Dorr, and K.T. Slattery (1998). "Low-Cost High-Speed Fiber Optic Grating Demodulation System for Monitoring Composite Structures", *SPIE Proceedings*, Vol. 3326, p. 390, 1998.

3. Udd, E. (1991). *Fiber Optic Smart Structures* in *Fiber Optic Sensors: An Introduction for Engineers and Scientists*, Edited by E. Udd, Wiley, 1991.

Temperature Compensation of Optical Fiber Voltage Sensor with dual light channels

Li Kaicheng   Feng Chunmei   Ye Miaoyuan

Huazhong University of Science and Technology,Wuhan ,430074,P.R,China

ABSTRACT

This paper introduces a new type of optical fiber voltage sensor with temperature compensation ability. The dual-light structure and operation of the sensor are discussed in great detail, and a series of test curves are given. The tests show that the dual-light channel voltage sensor has the ability of birefringence rejection and temperature compensation.

Keywords: optical fiber voltage sensor, dual light channels, temperature compensation

1.INTRODUCTION

Theory and practice show that the optical fiber voltage transformer has the problem of temperature stability, which seriously hamper their practical applications. The key of influencing the stability of optical fiber voltage transformer is the temperature birefringence of the crystal BGO($Bi_4Ge_3O_{12}$).It is due to the impurity and internal stress of the crystal. The birefringences are randomly distributed in the crystal and sensitive to temperature. When the temperature is changed, the output of the transformer will follow the change, which makes the measurement accuracy poor. Therefore, some measures must be taken to overcome the influence. Tests show that the dual-light channel method is a good way to improve stability of the sensors.

2.PRINCIPLE OF THE COMPENSATION

Fig.1 shows the structure of the voltage sensor with dual light channels. The optical process is as follows: A LED emits light with certain wavelength(850nm),then the light passes through a polarizer to be changed to circularly polarized light. After that, the quarterwave plate changes the linear polarization to circular polarization Under the action of electrical field, the light takes place birefringence to form elliptically polarized light. After a polarizing beam splitter (PBS,Wollaston prism),the light is split to

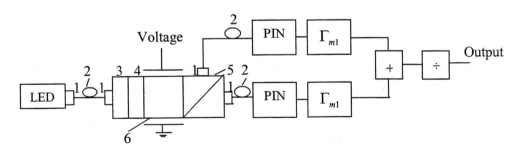

Fig.1 Scheme of dual -light channel voltage sensor
1. lens   2.optical fiber   3.polarizer   4.1/4 wave plate
5.polarizing beam splitter   6.electrooptical crystal

Correspondence:Email:likc@public.wh.hb.cn;Tel:86-27-87540573

Industrial Sensing Systems, Anbo Wang, Eric Udd, Editors,
Proceedings of SPIE Vol. 4202 (2000) © 2000 SPIE · 0277-786X/00/$15.00

two linearly polarized lights perpendicular each other, The reflected and transmitted signals from the PBS are equivalent to the signals transmitted from the crossed polarizer and from the parallel polarizer, respectively. The two light beams are respectively received by two PINs. According to the relation discussed in paper[1][2],the parallel modulation depth and perpendicular modulation depth respectively are:

$$\Gamma_{m1} = \frac{\Gamma_m}{1 - \sum_{n=1}^{N} \delta_{ml} \sin 2\theta_n} \tag{1}$$

$$\Gamma_{m2} = \frac{\Gamma_m}{1 + \sum_{n=1}^{N} \delta_{ml} \sin 2\theta_n} \tag{2}$$

where ,modulation depth $\Gamma_m = \frac{2\pi}{\lambda} n_0^3 \gamma_{41} V \frac{l}{d}$, n_0 is the reflective index of the crystal , γ_{41} is its electrooptic coefficient, l is the length of the crystal in the direction of light, d is the width of the crystal in the direction of the applied electrical field, λ is the wavelength of the light,V is the applied voltage. δ_{nl} is the phase retard caused by the n-th linear birefringence, θ_n is the slow axis angle. Γ_{m1} is the parallel modulation depth, Γ_{m2} is the perpendicular modulation depth.

Because $A = \sum_{n=1}^{N} \delta_{nl} \sin 2\theta_n$ is quite small, so Γ_{m1} and Γ_{m2} can be expanded as follows:

$$\Gamma_{m1} = \Gamma_m (1 + A + A^2 + A^3 + \cdots) \tag{3}$$

$$\Gamma_{m2} = \Gamma_m (1 - A + A^2 - A^3 + \cdots) \tag{4}$$

So, we have

$$\frac{\Gamma_{m1} + \Gamma_{m2}}{2} = \Gamma_m (1 + A^2 + A^4 + \cdots) \tag{5}$$

If the high order quality of A is ignored, we get

$$\frac{\Gamma_{m1} + \Gamma_{m2}}{2} \approx \Gamma_m \tag{6}$$

Γ_m is independent of temperature T, so the influence caused by temperature is rejected.

3.TEST RESULTS

The tests are aimed to examine the temperature compensation effect through using dual light channels. The method is as follows:

Keeping the input voltage 5kV unchanged, and altering the ambient temperature, we get the output voltage curves of the sensor with respect to temperature T.Fig.2 shows the test curves. In every picture, curve 1 stands for the parallel modulation depth, and curve 2 the perpendicular modulation depth, curve 3 the compensated resultant average modulation depth.The temperature

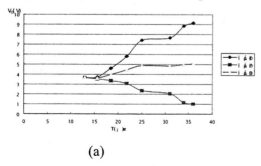

(a)

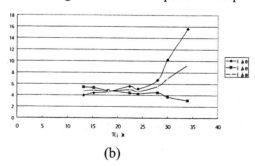

(b)

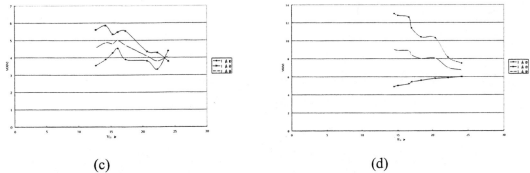

(c)	(d)

Fig.2 Temperature test curves

is changed from low to high and then from high to low, and enough long stabilized time is needed at any temperature point.

The above tests show that the modulation depths of the two light channels change oppositely as temperature rises, that is, one curve is up, and the other down. Similarly, the two modulation depths have the opposite changing tendency as temperature down. However, the average modulation depth is between them, and the curve seems flater.

4.CONCLUSIONS

A series of tests show that dual-light channel method is a effective way to eliminate the influence caused by temperature birefringence, and hence improve the stability of the sensor.

REFERENCES

1.Kyung S Lee, New compensation method for bulk optical sensor with multiple birefringences, Applied Optics,Vol.28,No.11,July,1989
2.Kyung S Lee, Electooptc voltage sensor: birefringence effects and compensation methods. Applied Optics.Vol.29,No.30,1990

Improved grating based temperature measurements using deconvolution of optical spectra

F. R. Ergul[a] ; T. Ertekin[b]
[a] Middle East Technical University, Electrical Engineering Department, Ankara, Turkey
[b] Aselsan AS, Balgat, Ankara, Turkey

ABSTRACT

The basic concepts of temperature measurement in power transformers and underground power cables using fiber optic gratings are introduced. The economic benefits of continuous temperature measurements are discussed. The problems encountered in high density applications, i.e. using many sensors in the same power transformer, is identified as the resolution limitation due to comparatively large bandwidth of optical filters used. The deconvolution technique has been proposed as a possible solution to improve accuracy and resolution. The fundamental ideas are illustrated by giving results of a laboratory prototype using twelve measurement points. The prototype has a capacity to employ up to 30 gratings for each phase of a power transformer, connected to the same measuring unit using an optical multiplexer.

Keywords: Power systems, transformers, cables, Bragg gratings, tunable filters, deconvolution.

1. INTRODUCTION

The local heating in transformers and underground cables, called hot spots, either pose a very serious threat to safe operation of power systems or avoid utilities operate their network under the optimum economic conditions. It is evident that transformer or cable failures are very costly and should be avoided at all cost. At the other hand, safe but sub-optimum operation of power system means reduced revenues. It also implies unused capacity, or simply overcapacity, in generation and distribution. Hence, economic benefits expected from a reliable and low cost technique to measure temperatures inside transformers and cables are tremendous. In this work, the emphasis will be on applications in transformers but similar economic considerations and possible applications are valid in the case of underground cables.

Temperature distribution in windings and at other points inside a transformer determines the loadability of transformer. The temperature profile between interiors and exteriors depends on environmental conditions such as wind, temperature, precipitation, solar heating and load types. The peak loads exceeding the name plate ratings, their magnitude, duration and frequency determine the rate of shortening of the insulation life.

For safer operation of transformers, utilities has the option to run computer simulations to determine loading characteristics. This may enable them to operate transformers by making best use of the capacity. Simulations under realistic conditions improve the adhoch knowledge acquired by utilities during many years of operation. However, if an insulation weakness develops as a result of stresses or aging, such modeling will not help in any way to guarantee safe operation. The only way to ensure safe operation is to know true values of interior temperatures of transformers.

The reasonable solution to the problem is continuous monitoring of internal temperature at critical points and transfer temperature distribution information to SCADA system for use in optimizing system operation and/or to take corrective or preventive measures in the system. Direct information on possible hot spots could improve loading practices and free utilities from strictly adhering to name plate data based loading.

The critical point here is that, operation on computer based simulation data can not avoid hot spot occurrence. When a hot spot occurs due to some reason, gas bubbles will start to develop in oil filled types. This is more serious then simple aging of insulation. In resin core transformers, similar dangers are present. Hot spots may cause quick deterioration of insulation which lead to dramatic drops in dielectric strength of the material.

Correspondence: F. R. Ergul, Email: ruyal@metu.edu.tr, Telephone: +90 312 210 2329, Fax: +90 312 210 2329

The benefits expected from constant monitoring transformer internal temperature profile could be summarized as follows:

1. Utilities can load transformers continuously at values larger then their nameplate values.
2. Under emergency conditions, transformers can be overloaded safely at higher load levels than accustomed for extended time durations.
3. Loading of transformers during normal operation and/or in case of an emergency can be done safely by choosing loading parameters according to variations in environmental conditions. For example, operating parameters during summer and winter may be selected entirely differently for the same transformer,
4. Possible causes of failures can be diagnosed early so that preventive actions can be taken and/or maintenance schedule can be altered.
5. By increasing the load carrying capacity of existing networks, new capacity investments can be eliminated or delayed.

There are some additional benefits expected from temperature monitoring of transformers if installation is done during the manufacturing cycle of the transformer at the factory.

1. Installation cost will be low.
2. Increased safety and operational advantages make transformers with such measurement capability more competitive in the market.
3. Cost, size and weight of the transformer may be reduced by using smaller safety factors in the design.

As it is known, there has been considerable research activity in United States and other countries to develop reliable and economically feasible techniques for temperature measurement inside power transformers and cables. Some of the references give an idea about the level of activity. However, it is our belief that still there is no satisfactory solution which has wide spread acceptance.

Compared to classical techniques based on thermo-couples and strain gauges; fiber based systems have several advantages. They are very suitable for applications in the hostile environment present inside a transformer, that is, strong electromagnetic fields, disturbing transients and noisy conditions. The fiber optic based solutions do not jeopardize the dielectric integrity of the transformer. There is no metallic part and the thin optical fiber assembly can be easily integrated into transformers internal windings. Temperatures will be measured at points known with accuracies given in centimeters. The inherently simple construction requires no maintenance and complete measurement system has long life.

Furthermore, fiber sensors and measurement system are connected to each other by using a fiber cable. Hence, their separation can be large and measurement unit and the sensors can be placed in different mediums. This is a property that makes it easy to take measurements in difficult to reach positions. One example of this could be power transformers located in switchyards.

Recognizing these advantages, work on fiber optic sensors for power system applications has started as early as 1991. The basic approach was to use optical fiber's temperature dependent properties for sensing. Such approaches are difficult to implement and cost was comparatively high.

Recent developments in optical communications, especially in the field of wavelength division multiplexing, has led to important developments in "tunable filter" and "in-fiber Bragg Grating" technology. Today, they can be purchased off the shelf at reasonable prices. With the development of wide band optical sources, such as ASE light sources, it has been possible to set up temperature measurement systems using multiple sensors with very good accuracies.

The sensor assembly could be a seamless single fiber (communication grade optical fiber) with ultraviolet printed in-fiber gratings at required positions. The length of the fiber would be selected according to the application. This sensing assembly is the only part that has to be placed inside the transformer. Other components of the system such as, 3dB coupler, isolator, in-fiber Bragg grating elements, source, detector and Fabry-Perot tunable filter can be placed at a distant location up to a few kilometers. They are all well proven long life, no maintenance devices. When combined with powerful signal processing capabilities of a laptop computer and its data acquisition means, they lead to practical and low cost measurement systems.

Therefore it is our conviction that Bragg grating based systems are very promising and cost effective alternatives to other known approaches in power transformer temperature measurements.

There are two issues which can be viewed as disadvantages in implementation.

1. The optical fiber is somewhat fragile for handling in manufacturing facilities of transformers. Hence, placing the fiber among windings require special procedures and very careful planning.
2. The fiber's protective coatings used should not contain any material which can decompose during operation inside the oil filled environment of transformers.

These two problems are investigated in great detail and it is the transformer and fiber manufacturer's opinion that they are manageable problems.

In this paper, a laboratory prototype of a measurement system will be introduced, the problems encountered will be discussed. Solutions to resolution problem which is due to relatively wide optical bandwidth of Fabry-Perrot filter will be presented.

2. EXPERIMENTAL SYSTEM

2.1 Introduction

The block diagram of the experimental system is given in Figure 2.1. A broadband optical source illuminates an array of twelve fiber Bragg gratings, each reflecting energy at its central wavelength, Wavelengths are separated from each other by a distance of 2 nanometers. The reflected wave passes through a 3dB coupler and the Fabry-Perot tunable filter. Energy is detected using an optical detector and electrical output is coupled to the input of data acquisition card of computer. The measured power is digitized by data acquisition card A/D converter and readings are transferred to the measurement software. The sampling rate is selected by software according to the span of wavelengths swept and the required accuracy.

The center wavelength of the filter is set by using 14-bit digital output of data acquisition card. This makes it possible to set filter wavelength with an accuracy of -/+ 0.0025 nanometers. The filter is swept between the desired start and stop wavelengths, at increments specified by software. Measurement results are stored as filter is swept. The power spectrum of the reflected energy is obtained by plotting power versus wavelength. The developed program processes the stored data and senses the peak points in spectra corresponding to maximum reflected power by each grating. Positional variations of these peak points indicate the amount of change in temperature relative to initial values. The amount of change is measured in nanometers and converted to degrees by the measurement program.

Since Fabry-Perot filter bandwidth is comparable to Bragg Grating bandwidth, there is a smoothing action which is significant. Its effect on the reflected power spectrum is to reduce accuracy and resolution of measurement.

This problem is overcome by using deconvolution of spectrum. Deconvolution algorithm is implemented by the signal processing part of the measurement program. In addition to the deconvolution, the developed software uses:

1. averaging and digital filtering to reduce noise effects,
2. peak detection to locate spectrum peaks.

Furthermore, software performs:

1. compensation of variations in power spectrum of the broadband source,
2. conversion of wavelengths to degrees centigrade to for each sensor,

All grating temperatures are displayed on the computer screen in a convenient way.

2.2 System Components

In this section, brief information on the hardware and software components used will be give. These components are data acquisition card, the Fabry-Perot tunable filter, fiber Bragg gratings, optical source, optical power detector, and a personnel computer.

2.2.1. Data Acquisition Card

It is a multifunction I/O Board type Lab-PC + from National Instruments. It contains a 12-bit successive-approximation ADC with eight multiplexed analog inputs, two 12-bit D/A converters, 24 TTL compatible digital I/O lines, and six 16-bit counter/timer channels for timing I/O. 14 lines of digital I/O used to set the center wavelength of the Fabry-Perot filter. The ADC onboard has capability of taking up to 1000 readings/second. Sampling instants are controlled by the timers on board.

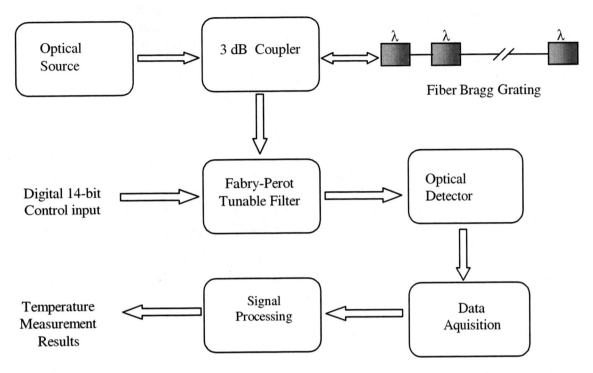

Figure 2.1. Block diagram of the measurement system

2.2.2. Fabry-Perot Tunable Filter

The reflected power from Fabry-Perot filter as a function of wave length has a typical shape as shown in Figure 2.2.

There are two important parameters that characterize the filter. Finesse, F is the parameter of the resonator which is a function of losses of mirror like surfaces. Typically F has values in the region 200 to 2000. The free spectral range, FSR is the distance between two peak points of the periodic spectra. It is dependent on structural properties of the filter and limited to values up to several tens of nanometers.

The reflected power I has sharp peaks if F is large. The peaks have a full width at half maximum (FWHM) given by (FSR)/ (F). That is, the width of each peak is F times smaller than the spacing between two peaks. The reflected power has a minimum which occurs at midpoint between two successive peaks.

If a Fabry-Perot etalon is used as a tuneable optical filter of a spectrum analyser, there are certain limitations and problems.

The spectral width of the measured signal must be narrower than the free spectral range. Otherwise, an ambiguity in measurement will arise.

1. The resolution of measurement is limited by finesse F of Fabry-Perot filter. Fine detail in spectra will be lost due to smoothing action of correlation.
2. There is a bandwidth expansion and resultant spectrum has a bandwidth larger than the individual bandwidths of signal and filter.
3. If FWHP values of an optical signal and the Fabry-Perot filter has values H_1 and H_2 respectively, the measured FPHW value will be approximately $H_1 + H_2$.
4. Fabry-Perot filter can not resolve two closely spaced narrow pulses if their FWHP values are small or comparable to Fabry-Perot FWHP value.

Therefore, successful application of Fabry-Perot filters requires special attention in measurement system design.

Tunable optical filter used is a DMF series device manufactured by Queensgate Instruments Ltd. UK. It operates in 1550 nm band. It has an FSR equal to 40 nanometers. The ratio of maximum value to minimum value occurring between two peaks is called Contrast and has value of 42 dB. The finesse is 200. The filter is electronically tunable via digital or analog interfaces with sub-millisecond response time. Fabry-Perot etalon mirror spacing and parallelism is controlled by piezoelectric (PZT) actuators and monitored by capacitance sensors, which operate in a closed-loop to stabilize both the cavity parameters. The capacitive sensors also linearize the wavelength scan when the filter is used in scanning mode. An out of range status indicator assures correct operation of filter. The 14 bit digital input is used to scan the 40 nanometerband in 16384 steps, providing a resolution of 0.0025 nm. The rest state value filter central wavelength is 1530 nanometers.

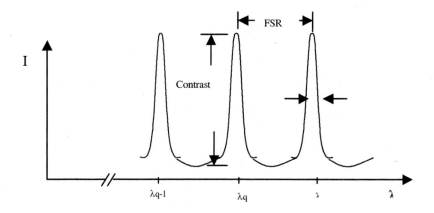

Figure 2.2 Fabry-Perot Filter Reflection Characteristics.

2.2.3. Bragg Gratings

Bragg gratings are devices mainly used as wave length selective filte rs in optical communication systems. They reflect light energy at their central wavelength and transmit energy with little loss at other wavelengths. This property makes it possible to form linear arrays of gratings by connecting them in series.

The intensity of reflected wave is a function of λ, the wavelength and Δs, the separation of slits that form the grating.

Gratings are formed by forcing refractive index variations in the core region of a communication grade optical fiber. The high index regions in the fiber core correspond to slits of a classical grating. The index variations are induced in the core region by laterally exposing fiber to UV light using high power ultraviolet lasers (for example eximer lasers), phase masks and specially designed computer controlled equipment. The automated manufacturing reduces the cost and increase the quality.

The sensitivity of reflected wave intensity on λ, enables one to use gratings as highly accurate strain and temperature sensors.

1. The strain applied to fiber along its axis causes Δs to increase. This variation will result in a change of reflected energy wavelength.

2. The sensitivity of core refractive index on temperature, causes changes in optical path length which are analogous to changes in Δs due to mechanical strain. Hence, temperature changes will be converted to changes in reflected energy wavelength.

The wave length change, which can be measured with accuracy, leads to accurate strain and temperature measurements. Hence, Bragg gratings are strong contenders in distributed temperature and strain measurements. They are expected to replace strain gages and thermocouples in demanding and critical applications.

For a specific Bragg grating [4], the following wavelength shift expression can be obtained:

$$\Delta\lambda/\lambda = 0.79\varepsilon + \xi\Delta T.$$

Here, ε is axial strain in the optical fiber core and ξ is the thermo-optic coefficient of the fiber. The constant ξ determines the thermal sensitivity is measured to be 6.3×10^{-6} / ^{0}C for this specific grating. This low thermal sensitivity correlates to a low thermo-optic coefficient of 0.01nm / ^{0}C around the 1550nm band. Therefore, a 100^{0}C change causes a wavelength change of 1nm. Since $\Delta\lambda$ can be measured with accuracies in the range of 0.01 nanometers, temperature measurements with degree accuracy is possible.

A typical Bragg grating reflection characteristics is given in Figure 2.3. It has an half power width of approximately 0.1 nanometer.

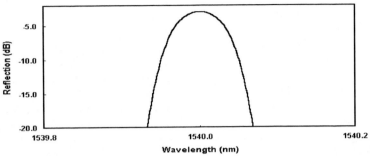

Figure 2.3. A typical reflection spectrum of a Bragg grating

The gratings used in our system are produced by Kromafibre, Canada. Twelve gratings are used in the setup. The nominal wavelengths of the gratings are from 1534 nm to 1556 nm with 2nm spacing. They have 98% reflectivity and the wavelength accuracy is ± 0.05 nm at 25 ^{0}C and zero strain. The temperature dependence is approximately 0.008 nm / ^{0}C.

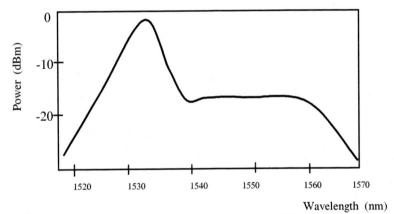

Figure 2.4. Output Spectrum of ASE Light Source

2.2.4. Optical Source

A wide band sources has to be used to illuminate all gratings simultaneously. It is necessary to use bandwidth larger than the measurement bandwidth, which is the difference of highest and lowest grating wavelengths. There are two optical sources which can be used for this purpose. They are edge emitting diodes and ASE (Amplified Spontaneous Emission) sources. ASE source has been chosen in the system for being a broadband source with a high output power of 10 miliwatts. The ASE source is manufactured by Lightwave Technology USA.. It has a highly stabilized, wide spectral of 40 nanometers centered in the 1550 nm telecommunications band. It is very suitable for using in conjunction with an optical spectrum analyzer for rapid, wide dynamic range spectral characterization of fiber-optic components such as filters and fiber Bragg gratings. Its output spectrum is not constant. Its variation with wavelength is shown in Figure.2.4.

2.2.5. Optical Power Detector

The power detector used in the system is HP 81532A Lightwave Multimeter from Agilent Technologies, USA. It is used in the system because of the very high flexibility it offers in measurements. Its programming features helped to obtain long term data easily.

2.2.6. Personal Computer

The computer used has a Pentium-166 processor. The LabVIEW program and the associated DAQ driver programs run under Windows. Since sampling rate is not very high, no capacity problem was experienced during experiments.

2.2.7. Application Software

The measurement system uses the software LabVIEW from National Instruments. The reasons of selecting LabVIEW for the system implementation are :

1. It is well suited for instrumentation applications.
2. Ease and flexibility in data acquisition and processing.
3. User-friend output screens for displaying measurement results

It enables the user to write signal processing modules easily and provides means for setting the start and stop wavelengths of the scan, selecting the sweep rate and increments. It also provides means to take a desired number of averages, to normalize source power, to perform digital filtering. The digital filtering implemented is a low pass FIR filter which is equi-ripple type with 99 taps. One the critical tasks performed by the software is the implementation of deconvolution algorithm.

The signal processing done on the measured spectra includes an algorithm for accurate location of peaks of spectra. This is necessary because small distortions in waveform leads to a small shifts like perturbations. Hence, such a curve fitting process reduces the errors introduced due to such distortions. He algorithm fits a quadratic polynomial to sequential groups of data points. The number of data points used in the fit can be changed according to the scan parameters.. During peak detection, obtained results are tested against a threshold level and peaks below threshold are ignored.

All application software modules are developed as VI's in Labview environment.

2.3. Experimental Results

The complete measurement setup and procedure has been tested in the laboratory. The results were in parallel with expectations. Due to limited resources, grating temperatures were not changed independently. Rather, one grating temperature was controlled and the others were kept at ambient temperature.

It was observed that the effective of 0.2 nanometers bandwidth of tunable filter did correspond to an equivalent bandwidth of 0.06 nanometers after deconvolution operation. This means, gratings can be placed more closely to each other and the number of gratings can be as much as three times as the case of direct measurements.

Using a simple oven, a series of temperature measurements were doe to determine the temperature coefficient of gratings and test their linearity. (Figure 2.5). The temperature coefficient of particular grating was found to be 0.007 nanometer/ degree centigrade. Due to noise present in the system and due to other error sources, such as distortion introduced during signal processing, the systematic error was found to be in the range -/+ 0.02 nanometers. This corresponds a temperature error of -/+ 3 degrees.

Table 2.1. Grating Temperature vs Measured Wavelength

Grating Temperature (^{0}C)	Wavelength Measured (nm)
21	1535.73
28.2	1535.79
34.7	1535.86
42.2	1535.92
49	1535.97

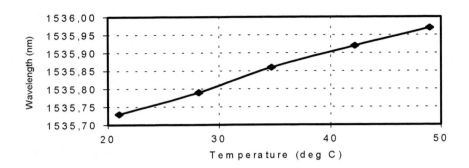

Figure 2.5. Grating Temperature vs Grating Peak Wavelength

The results are summarized in Table 2.1 and Figure 2.5. The deviations from linear behavior was difficult to interpret. It may be simply due to inaccuracies in oven temperature and grating temperature measurements. However, by performing a well designed calibration procedure using more sensitive temperature control mechanisms, such effects can be eliminated. Still, linearity seems to be satisfactory for most applications.

The appearance of user screen is illustrated in the sample screen shown in Figure 2.6.

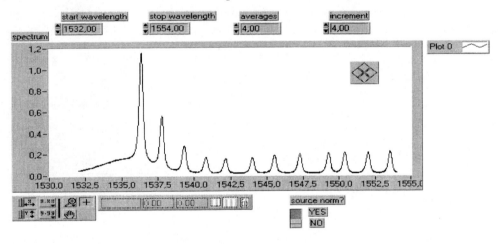

Figure 2.6. A sample screen representing temperature measurement by 12 gratings.

3. DECONVOLUTION PROBLEM

3.1. Introduction

If an signal s(t) is applied to input of a linear time-invariant system, with an impulse response h(t), the system's response y(t), due to imposed excitation is given by convolution

$$y(t) = \int_{-\infty}^{+\infty} s(\tau)h(t-\tau)d\tau = \int_{-\infty}^{+\infty} s(t-\tau)h(\tau)d\tau$$

When frequency domain notation is used, the same operation is represented by

$$Y(jw) = S(jw).H(jw)$$

These two equations are related to each other through Fourier transform.

The inverse process of the convolution operation is known as *deconvolution* Unlike most operations, the deconvolution operation has no direct mathematical definition. In frequency domain, the deconvolution operation can be represented as

$$S(jw) = Y(jw)/H(jw)$$

It is not difficult to assume that the division operation would create problems. It will lead to noise amplification in regions where H(jw) is zero or small. Without going into detail, one can argue that this problem is an estimation problem which is sometimes referred as an "ill-conditioned" or "ill-posed" problem. For this class of problems, more than one approximate solution may be achieved, some of which can be considered as "acceptable" estimates to the exact solution.

The convolution will be considered in frequency domain in the temperature measurement system[+]. The aim of deconvolution is to recover the frequency content mainly at high frequencies which was attenuated during the convolution.

3.2. Spectral Measurement Using Tunable Filters

The spectral measurement is done by using the tunable filter and a power measurement device. The filter center wavelenght is first adjusted to a point outside of the signal wavelength bandwidth. Then, filter is swept over a wavelength range larger then the combined signal bandwidthand and filter bandwidth. The power measurement device output, when plotted versus wavelength, gives the measured spectra. Mathematically, it is given by the equation:

$$Y(\lambda) = \int_{-\infty}^{+\infty} S(\tau).H(\lambda - \tau)d\tau = \int_{-\infty}^{+\infty} S(\lambda - \tau)H(\tau)d\tau$$

This equation represents the convolution in wavelength (or frequency) domain. The H(λ) is the window function in classical spectral measurement terminology. In the measurement setup under investigation, the bandwidth of tunable filter and bandwith of reflected energy from Bragg gratings are comparable in magnitude. (Figure 3.1).

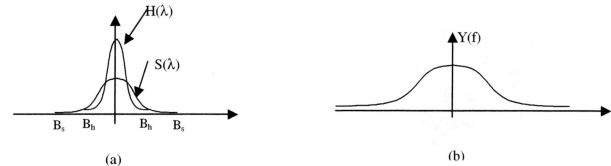

(a)   (b)

Figure 3.1. Changing shape of the signal spectra due to convolution.

This implies that:

1. Resultant bandwidth of measured spectra will be larger than the bandwidth of S(λ).

2. There will be a smoothing action on S(λ) so that fine detail will be lost.

The figure shown in Figure 3.2 illustrates this case. The undesirable result of the convolution is that two closely placed peaks in the signal spectrum will be difficult to distinguish. This effect is more pronounced when closely spaced peaks have un-equal magnitudes.

In principle, it should be possible to use deconvolution in wavelength domain to recover the lost fine detail in the measured spectra. In other words, deconvolution may be viewed as a power spectrum restoration process. However, the presence of noise limits the generality of these statements and its presence places severe constraints on the total restoration of the spectrum.

[+] In this paper, wavelength and frequency will be used interchangeably since the bandwidths are narrow and two sided filter spectrums are symmetric.

It can be shown that satisfactory results could be achieved with deconvolution if signal to noise ratio at the output is very high. The signal to noise ratio can be improved by filtering or smoothing of the spectral data. Always there is a practical limit to the amount of filtering or smoothing that makes sense. They effect the power spectrum in such a way as to attenuate the spectral power density at higher frequencies. If higher frequencies are attenuated in the power spectrum, it causes loss of information corresponding to high resolution Therefore, it is correct to say that deconvolution and smoothing are two processes with conflicting objectives. These arguments imply that when applying spectrum measurement techniques, care should be exercised and these points are taken into account.

3.3. Deconvolution Techniques

Deconvolution is a broad problem that spans numerous disciplines, and as a result there exists a rather rich literature on the subject. A major problem results from the equally rich sets of solution methods. Some of the important ones in the literature are Frequency Domain Methods [4], Iterative Nonlinear Methods [5], Cepstrum Analysis [5], Maximum Likelihood Deconvolution [6], and Maximum Entropy Deconvolution [7]. In the study reported, Frequency Domain and Iterative Nonlinear Methods which are devoted exclusively to spectrum analysis, were considered. Some of the promising techniques will be mentioned without giving any detail.

The Linear Deconvolution Methods try to find an inverse filter that could undo the convolution induced smearing in the signal. Since H(w) magnitude is small or zero for some high values of w, noise wil l be dominant and hence, a noisy and unusable result is obtained. In order to solve this noise problem at high frequencies, Wiener-Type Filters, Least Squares Deconvolution or Optimal Compensation Deconvolution can be used. They all try to find improved filter that can be used in deconvolution.

In Nonlinear Time Domain Deconvolution Methods, unlike the linear methods mentioned above, a correction on Y (either multiplicative or additive)is sought to improve the solution. Among the numerous approaches, Van Cittert's Method, Gold's Ratio Method, Truncation Method and Janson's Method can be mentioned.

The listed approaches were tested in reference [8] and it has been shown that Jansen's approach and frequency domain deconvolution both produces acceptable results.

In order to test and compare methods mentioned, an artificial object data spectrum, , with sharp characteristics is created. As Figure 3.2 shows, $S(\lambda)$ consists of two closely spaced pulses. The signal has been convolved with the transfer function shown in the same figure. The shape of the spectrum after convolution is as shown in Figure 3.3.The proposed deconvolution algorithms were then applied to recover the original signal spectrum through deconvolution. At this step of the study, same computer environment is used for implementing deconvolution (that is measurement setup computer and LabView program). Two VI's are developed for each method and tested. The results of 2 cases representing 10dB and 100 dB SNR cases were obtained.

For frequency domain deconvolution, the program first calculates Y[k] by using 1024 point FFT. H[k] is obtained the same way. Then, transform of $S(\lambda)$, S(k) is calculated using relation

$$S(k) = Y(k)\, H(k) / (\left| H(k) \right|^2 + \chi)$$

After several trial and errors, it was found that a convenient value for χ is 0.. $S(\lambda)$ is recovered by taking the 1024 point inverse FFT.

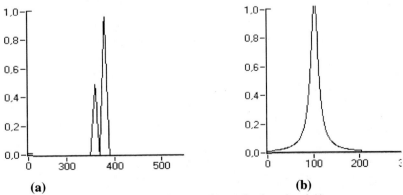

(a)   (b)

Figure 3.2. Sample signal s(t) and transfer function h(t)

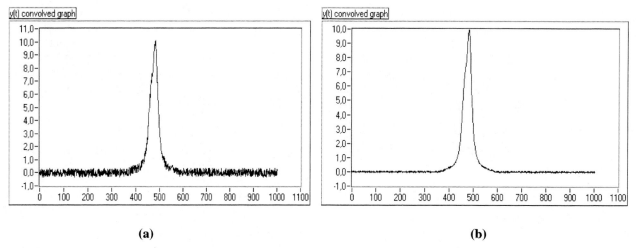

(a)   (b)

Figure 3.3. (a) Convolved signal with SNR = 10. (b) Convolved signal with SNR = 100.

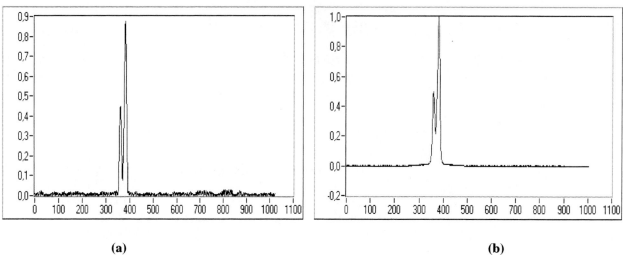

(a)   (b)

Figure 3.4. Deconvolved Spectra for SNR=100, (a) frequency deconvolution, (b) Jansson's method,

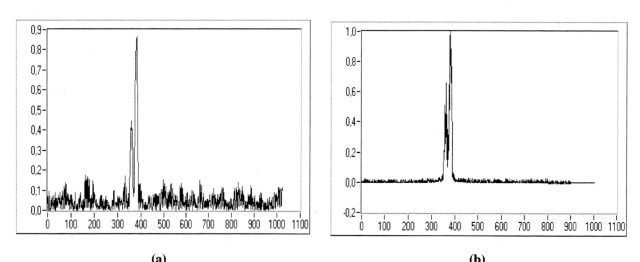

(a)   (b)

Figure 3.5. Deconvolved Spectra for SNR=10, (a) frequency deconvolution , (b) Jansson's method,

The program which implements Janson's algorithm uses a relaxation function. For being more efficiently, first 20 iterations are performed with $r_0 = 1$, and next 40 iterations with $r_0 = 2$. $Y(\lambda)$ is normalized to 0.5, as required, and the output estimate is normalized to 1, which is the peak value of the input signal. The results obtained using both methods for SNR = 100 are in Figure 3.4.a-b. The estimates of both methods for SNR = 10 are in Figure 3.5.a-b.

The software determines the location of peaks for each deconvolution procedure. Using frequency domain approach, determined values are exact, equal to 360 and 380 respectively. Janson's method, at the other hand, finds peaks at locations 361 and 379, making a slight error in values. As far as accuracy concerned, performances can be considered the same.

However, frequency convolution results in higher noise at off-peak regions where Janson's technique produces less noise term. When shape is examined around peak points, frequency domain approach has a cleaner look around peak points compared to Janson's technique results.

Because of the iterative nature of non-linear techniques, Janson's approach takes more time to process data. Perhaps this is a disadvantage if number of sensors are increased during temperature sensing operations.

4. RESULTS AND CONCLUSIONS

In the laboratory environment, the developed system was fully tested. It was observed during laboratory testing that the developed approach was rather sound and can be used in reliable temperature measurements in hostile environments such as power transformers and underground cables. Special signal processing tools which were developed to increase the accuracy of measurements also helps to combat against noise and remove the undesirable effects of convolution during spectrum measurement operation.

The developed measurement setup is being modified so that it will use a laptop computer and specially developed source and detector modules. When these units are completed, measurement unit will become an industrial distributed temperature sensing unit. The estimated number of sensor that can be used is large. It is expected that, using super luminescent diodes, despite their narrower power spectra, one optical fiber can contain up to 30 sensors. That means, if a separate fiber is installed for each phase of a transformer, a total number of 90 sensors can be employed This number seems to be sufficient for transformer applications.

The intensive investigations on how to place fibers in transformers indicate that using separate fibers for each phase is necessary. The optimum strategy would be to bring the fibers corresponding to each phase out from transformer casing and connect to measurement unit employing connection boxes mounted externally to transformer.

These estimates are based on the assumption that temperature variation can be 100 degrees centigrade. If it can be assured that less temperature differences can be expected, such as the case of underground cables, then the number of gratings that can be used per fiber increases. The total depend on measurement range but it is safe to predict that this number could be in the rage 50 to 60.

The transformer manufacturers in Turkey has continuing interest in the project and future work will be concentrated on developing a compact industrial measurement system specifically developed for power transformer applications.

References:
1. James, S.W.; Dockney, M.L.; Tatam, R.P.*Simultaneous independent temperature and strain measurement using in-fibre Bragg grating sensors,* , Electronics Letters v 32 n 12 Jun 6 1996. p 1133-1134: 1996
2. Boiarski, A.A.; Pilate, G.; Fink, T.; Nilsson, N,*Temperature measurements in power plant equipment using distributed fiber optic sensing,* IEEE Transactions on Power Delivery v 10 n 4 Oct 1995. p 1771 -1778,: 1995
3. Abdelhak Bennia, and Sedki M. Riad, *An Optimization Technique for Iterative Frequency –Domain Deconvolution,* IEEE Transactions on Instrumentation and Measurement , Vol 39, No 2, April 1990.
4. Sedki M.Riad I, *The Deconvolution Problem :An Overview,* Proceedings of the IEEE, Vol.74, No 1, January 1986.
5. Jansson, Peter A., *Deconvolution with Applications in Spectroscopy,* 1984, Academic Press.
6. Technical Review, Cepstrum Analysis, No 3, 1981, Bruel&Kjaer Instruments. .
7. Frieden, B.R.,1983, Probability, Statistical Optics and Data Testing, Springer-Verlag, New-York.
8. T. Ertekin, *Wavelength Measuring Techniques Using Fabry-Perot Tunable Filters*, M.Sc. Thesis, Middle East Technical University, December 1999.
9. Saravolac, M.P,*Use of optic fibres for temperature monitoring in power transformers* , IEE-Colloquium (Digest) n 075 Mar 22 1994. Published by IEE, Michael Faraday House, Stevenage, Engl. p 7/1-7/3, 1994.

A fiber optic remote inspecting technique for caverned large oil tanks

Weilai Li[a], Desheng Jiang[a], He Cao[b]

[a]Fiber Optic Sensing Research Center, Wuhan University of Technology,
122 Luoshi Road, Wuhan, 430070, China
[b]Dept. of Physics, University of Rhode Island, Rhode Island, 02881, USA

ABSTRACT

In the management of caverned fuel oil inventory, a strict rule of fire control has always been the first priority due to the special conditions. It is always a challenge to perform automatic measurement by means of conventional electrical devices for inspecting oil tank level there. Introduced in this paper is a fiber optic gauging technique with millimeter precision for automatic measurement in caverned tanks. Instead of using any electrical device, it uses optical encoders and optical fibers for converting and transmitting signals. Its principle, specifications, installation and applications are discussed in detail. Theoretical analysis of the factors affecting its accuracy, stability, and special procedures adopted in the installation of the fiber optic gauge are also discussed.

Keywords: optical fiber, oil, level, tank, cavern

1. INTRODUCTION

Fuel oil is one of the most important strategic materials. People seriously concern its inventory conditions, taking its security as the first priority. Some large oil tanks are even built in caverns or underground. The advantages to store fuel oil in caverns are, for instance, stable temperature, less evaporation loss and security.

Because the ambient temperature in caverns is stable (around 15°C), the density of all kinds of fuel oil in that temperature can be accurately calculated, and so are the dimensions of the tank. Therefore the question remained for automatic inventory management is measuring the exact level. Taking a 10,000m^3 gasoline tank for instance, if a millimeter of level is mistaken, the error would be as large as 600 kg (225 gal) gasoline.

However, the problem of caverned oil storage is its poor ventilation condition, the inflammable gas vaporized from the tanks is difficult to vent out. Therefore, in cavern tank bank, rules of fire control are strictly followed, even electrical lighting facilities are not allowed. Under this circumstance, it is difficult to use the conventional electrical/electronic instruments for performing automatic level measurement. Workers have to use a band tape to manually measure oil levels.

The fiber optic sensing (FOS) level gauge we have developed for caverned oil tanks is a good substitute measurement method. It features high accuracy, high reliability, and immunity of electromagnetic interference. Of most importance, it uses light to monitor the level without any electricity in the tank bank.

2. PRINCIPLE

In our current design, we use a float to directly measure the change in oil level. Compared with indirect measurements, it is more accurate to measure the physical quantity directly. Indirect measurement such as measuring the pressure on the bottom of tank and converting it into level value introduces unpredictable errors.

The FOS oil level gauge is composed of a mechanical driving unit, an optical fiber encoding unit and an electronic processing/displaying unit, as shown in Fig. 1. The optical fiber plays the roles of converting and transmitting the oil level signal. This type of FOS gauge can be considered belonging to the category of transmitting FOS.[1]

Industrial Sensing Systems, Anbo Wang, Eric Udd, Editors,
Proceedings of SPIE Vol. 4202 (2000) © 2000 SPIE · 0277-786X/00/$15.00

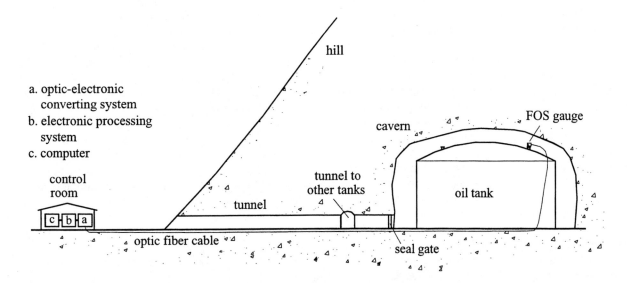

Fig. 1. A FOS oil level gauge used in a caverned tank.

a. optic-electronic converting system
b. electronic processing system
c. computer

control room

hill

cavern

FOS gauge

oil tank

tunnel to other tanks

tunnel

optic fiber cable

seal gate

2.1. The mechanical driving unit

The mechanical driving unit converts the change in oil level into rotation of the optical fiber encoder via the linear displacement of the float. Two pieces of mini-stainless steel rope and two coaxial spiral rope drums are used, as shown in Fig. 2. When the level is still, this unit is in equilibrium. The balance force is from the gravity of the balance weight.

When the oil level is up or down, the tension of the float rope drives the system to a new equilibrium. Therefore, a small change of oil level causes a corresponding change of circular angle of the drum. By means of a magnetic coupling device, the fiber encoder synchronically revolves with the float drum.

The float rope and the weight rope are wound in their rope drums respectively. Since the diameter of float drum is 4 times as large as that of the weight, when the oil level moves 4 meters, the weight moves only 1 meter. This gearless method of speed change is proved to be simple and effective. It uses a shorter weight guide pipe compared with a mechanism with no speed change, an advantage for installing the FOS gauge in caverns where the space above the tank is limited.

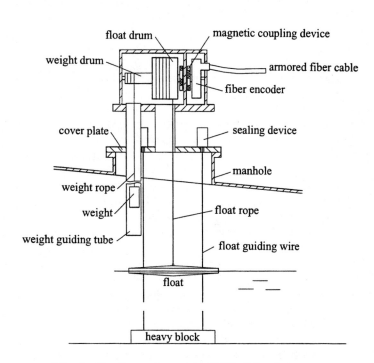

Fig. 2. Configuration of the FOS oil level gauge.

2.2. The optical fiber encoding unit

This unit is composed of a magnetic coupling device, two pairs of optical fibers and collimating lens, and an encoding disk with teeth around it. All these are assembled in the encoder compartment.

In summer, moist air condensed on cavern walls and tank walls deteriorates the working condition for the optical encoder, which needs a clean and dry condition. A patented magnetic coupling device is used to provide the encoding unit a well-sealed environment, water vapor or oil vapor are kept away from the optical encoder, while the rotary motion of the drums can be precisely transmitted to the encoder without direct contact.

The configuration of fiber encoding device is shown in Fig. 3. The continuous light output is from the end of left fiber, where located the focal point of the left focusing lens. This very small section can be considered as a point light source, so a parallel beam is formed on the right side of the left lens. Then the light beam passes through the right focusing lens, focuses at the right focal point and coupled into the right fiber. When the encoding disk rotates, the light beam is blocked intermittently. The receiving fiber gets a train of pulse. A pair of such fiber optic chopper (A and B, shown in Fig. 4) is necessary for making a directional counter.

For reasons of low cost and convenience for assembling, double convex lenses (DCX) with 8 mm effective focal length and 8 mm diameter are used. The fiber is $\phi\,62.5\,\mu m$ in core, multimode, silicon based. Its numerical aperture (NA) is 0.275. Making use of the definition of NA[3],

$$NA = n_0 \sin\phi,$$

the emitting angle ϕ of the fiber is 15.7°. The diameter of collimated light beam between two lenses is about 4.5 mm.

The circumference of encoding disk is 800 mm with 200 teeth, and each groove or tooth is 2 mm wide. If 4.5 mm light beam is used, the teeth cannot totally block the light. We put a slice (shown in Fig. 5) with a hole in the center to limit the light beam diameter. The light power loss caused by this optic stop can be easily compensated by using right light sources and photo-detectors.

With no slices, the transmitting efficiency has a maximum, 40% (-4 dB). The sources of the loss are mainly the spherical aberration of the lens, the reflection loss of the uncoated lens[2], the coupling deviation of lenses and the centric deviation of two fiber end sections. The relation between light power loss and the diameter of covering hole is illustrates as Fig. 6. The experiment shows that choosing 1.4 mm diameter hole of covering slice, with –10 dB transmitting loss, can get a good result in both the shape of output light pulse and the power of light.

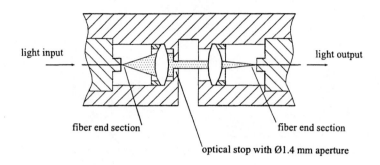

Fig. 3. Construction of fiber encoding device.

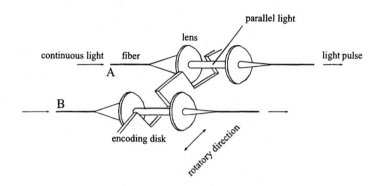

Fig. 4. Configuration of fiber encoding device.

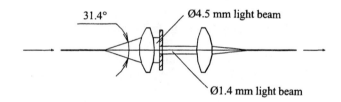

Fig. 5. Stop for limiting the beam size.

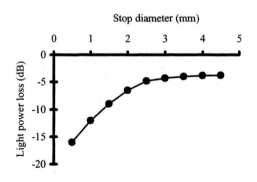

Fig. 6. Relation between power loss and stop diameter.

When the encoding disk rotates a full circle, two groups of 200 periods of pulse output and they are superimposed. The period difference between them is set as 1/4 of a period, referring to Fig. 4. So there are 800 edges in two groups of pulse, each edge can trigger the counter, and one edge represents one millimeter of level change.[1]

2.3. The photoelectrical converting unit

The fiber cable brings the light pulse signal into control room. In this unit, two pieces of 850 nm LEDs (HFBR-1424, HP) are used as light sources. Two PIN detectors are used to convert light pulse into electric pulse.

The two electric pulse trains are amplified and conditioned in this unit. In order to strengthen the reliability, the light threshold of triggering value is adjusted so low that the actual value of light pulse is thousands of times more than the threshold value, for adapting the attenuation of LED output after long term use.

Fig. 7 is a diagram of optic-electronic converting unit. The PIN detectors convert the light pulses into current signals. After the current signal is converted into voltage signal for comparison with a preset threshold value, an Schmitt trigger is used for signal conditioning. When the encoding disk rotates, it blocks the light beam gradually. The output wave graph, in Fig. 7, shows that instead of steep edges, the transit time is too long (left). After Schmitt trigger, the signal becomes a square wave (right), good for triggering the counter circuitry.

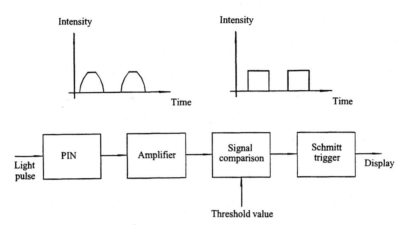

Fig. 7. Photoelectric converter.

2.4. The processing unit

Shown in Fig. 8 is the instrumentation using a single-chip microcomputer. The CPU is a quasi-16 bit micro-controller, 80C196 (INTEL Corp.). It can handle not only programmed calculation, but also the calibration, self-compensation and self-adjustment.[4] This unit performs direction determining, counting, compensating, and displaying. According to the input signal pulses monitored, the corresponding fuel level is displayed with two methods, one is in millimeter digits and the other is an analogue display of a column of LEDs.

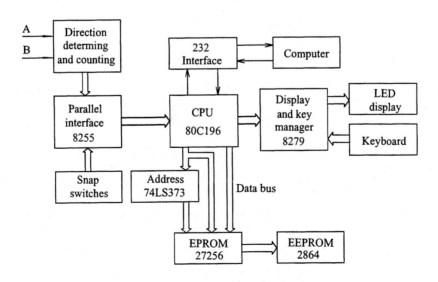

Fig. 8. Data processing unit.

3. INSTALLATION

One of advantage of FOS oil gauge is that it can be installed when the tank is still in operation, thus eliminating the need to dump the remaining oil. Installing on-line means that none welding, drilling or electric tool is used. This is a great economical benefit, because the preventive measure of fire control cost too much. This is also one of the reasons that conventional electronic instruments cannot be installed in caverns whose installations always interrupt the tank operation.

A unique design of FOS gauge has been developed, illustrated as Fig. 2. The installation procedures are as follows:
- Find a manhole on the top of a tank, and take its cover plate out off the cavern and machine it for preparing the installation.
- Carefully sink a heavy steel block to the bottom of the tank, tie two pieces of stainless steel wire as the guide-bar of the float on the block. It keeps the float from being drifted away by the oil current (if any). This construction gives the float an ideal working condition for the measurement.
- Put the float into the tank and recovery the cover plate and fix the wires.
- Install the gauge unit and optical fiber cables.
- Calibration and programming.

4. AFFECTING FACTORS OF ACCURACY AND COMPENSATION

4.1. Diameter of float

The moving system moves step by step, and the larger the diameter of float is, the higher the accuracy it can be.[1] A float with 480mm diameter is used, which is just a little smaller than the size of man-hole, 500 mm in diameter. Taking diesel oil for example, when the oil level goes up, the discrete moving step of the float Δh_F is 0.13 mm, when the oil level goes down, Δh_F is 0.08 mm.

4.2. Ambient temperature

Because of the difference between the coefficient of thermal expansion of stainless steel rope and the structure steel of the tank wall, the measuring error caused by temperature change Δh_T can be written as:[5]

$$\Delta h_T = (d_1 - d_2)(T_S - T_W)L,$$

where d_1 is the coefficient of linear expansion of stainless steel; d_2 is the coefficient of linear expansion of structure steel; T_S is temperature in summer; T_W is temperature in winter; and L is the level range. Taking $d_1 = 1.6 \times 10^{-5}$ m, $d_2 = 1.0 \times 10^{-5}$ m, $T_S = 15°C$, $T_W = 14°C$, and $L = 15$ m, then Δh_T is 0.09 mm.

4.3. Position of measuring rope

From the specification of the stainless steel rope, the rope construction is 0.9 mm in diameter, 7×7 strands with self-weight 0.032 N/m and breaking tensile strength higher than 539.0 N. We can figure out that when the float is at the bottom point, 12.5 meters long rope give the float 0.40 N additional weight, and it makes float sunken Δh_{R1} 0.27 mm. When the float is at the top position, the rope is only a quarter of 12.5 m long which adds the weight 0.10 N, and because of the 1:4 speed change, it makes float upward only 0.02 mm.

The tolerance in diameter error of the rope is ± 0.025 mm, according to its manufacturer. If the radius differs in 0.025mm for a designed 800 mm circumference, the error in that circumference is 0.15 mm. 12.5 meters long stainless steel rope twisted on the spiral drum is about 16 circles, the accumulative error Δh_{R2} can reach to 2.51 mm

Other factors such as the increased rotary inertia when the 12.5 m rope twists on the drum, and the additional moving resistant force that also affects the measuring accuracy, are ignored.

4.4. Processing error of float drum

If the processing error in 800 mm circumference is 0.1 mm, the rate of error is only 0.0125%, however, it will cause an accumulative error Δh_P, 1.6 mm for 16 circles. Therefore, a precise float drum is needed. In manufacture process, precision finishing costs too much. At first, we made the drum's circumference to be adjustable in a small range. But this increases the cost by many incremental parts and workload and was not adopted.

4.5. Compensation

With above considerations, we find that the total error, $\Delta h = \Delta h_F + \Delta h_T + \Delta h_{R1} + \Delta h_{R2} + \Delta h_P = 4.60mm$, is bigger than the resolution, 1 mm. The solution we have adopted is to compensate measuring data by the data processing unit.

First we put a gauge in a test tank, operating it from zero level to full range. At the full level, we get two values, a value from inside counter M_f and a true value T, which is from a tape measure by the qualified measuring technicians. Secondly, input T into the EPROM of processing unit. The assembler-language program gives CPU the command for calculating $C = 1 + M_f / T$, where C is compensation value, as shown in Fig. 9.

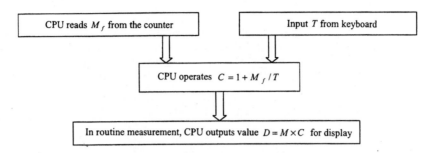

Fig. 9. Compensation algorithm.

C may be either larger than 1 or smaller than 1, but the deviation is very small 1.0005 or 0.9996, for instance. When the FOS level gauge gets a counter datum M, the CPU multiplies M by C, and then displays the product. The product keeps only integers. For example, if the compensation value C of a FOS level gauge is 1.0005, when M is less than 2000 mm, there is no need for compensation at this moment. When it reaches at 2000 mm, the processing unit displays 2001 mm, and in every next 2000 mm, the CPU adds 1mm to display, and vice versa. With this method, the display value is exactly the same as the true value in every point of tank range.

As a measuring instrument with ±1 mm accuracy in every 10-meter range, this oil FOS gauge system has been tested and certified by (Chinese) National Technology Supervision Bureau.

5. DISCUSSION

We have paid great attention to the reliability of FOS gauge operation including float movement and optical fiber encoder to ensure the high precision of the FOS gauge system. Because with a reliable measuring and encoding system is the base of compensation. The practical usage has proved that this technique is successful. It is an effective technique to replace the outdated measuring method for the management of cavern tank banks.

REFERENCES

1. W. Li, D. Jiang, and H. Cao, "FOS gauging technique of Yangtze River level with millimeter accuracy," *SPIE Proc. 3538*, pp295-303, 1998.
2. J. Xu, *The fundamentals of engineering optics*, 3rd edition, Press of Machinery Industry, Beijing, 1994.
3. D. Liu, and Z. Yin, *Fiber Optics*, Science Press, Beijing, 1987.
4. Y. Zhang, and B. Chen, *Practice Handbook of MCS-8098*, 2nd edition, Tsinghua University Press, Beijing, 1994.
5. W. Wang, *Practice Handbook of Mechanical Design*, Press of Agricultural Machinery Industry, Beijing, 1985.

Addendum

The following papers were announced for publication in this proceedings but have been withdrawn or are unavailable.

[4202-03] **Fiber optic sensors for automotive engine control**
D. Barua, B. Davis, Visteon Corp. (USA)

[4202-04] **Long-life fiber-optic-based pressure sensors for combustion engine monitoring and control**
O. E. Ulrich, M. T. Wlodarczyk, OPTRAND, Inc. (USA)

[4202-05] **Fiber-optic-sensors-based optonumerical relay**
N. Khan, C. G. Yung, Univ. Putra of Malaysia

[4202-06] **Development of an optronic over-current relay**
N. Khan, I. Aris, Univ. Putra of Malaysia

[4202-09] **Optoelectronic system for control of fuel ignition in internal combustion engines**
M. M. Kugeiko, A. Scherbatyuk, Belarussian State Univ.

[4202-10] **Pure photonic vehicle management system architecture**
T. L. Weaver, Boeing Co. (USA)

[4202-20] **Sensing and repair of bridge transverse cracking and internal shear stress/ fracture and of failure of prestressed girders or prestressed deck plates**
C. M. Dry, Univ. of Illinois/Urbana-Champaign (USA)

[4202-23] **Advanced monitoring and characterization of coal-fired flames**
Y. Yan, G. Lu, A. R. Reed, Univ. of Greenwich (UK)

[4202-26] **Fiber optic current sensors for application in large power generators**
T. Bosselmann, S. Hain, M. Willsch, Siemens AG (Germany)

Author Index